AI FAIRNESS AND BEYOND

This book proposes a regulatory system for ensuring that AI makes fair decisions.

No one wants to be the subject of an unfair decision made by an AI, and fairness is so important to society that we are likely to want to regulate to demand it. But how? This book attempts to answer that question.

The aim of regulation must be for an AI's decisions to match the human conception of fairness. To understand what that is, the book proposes a holistic understanding of fairness, which tells us what regulation must try to achieve.

However, regulation is not an abstract activity – it regulates how humans behave, and the humans in question are those who develop and use AI for decision-making. Thus the book investigates how those humans are attempting to achieve AI fairness. It finds that there is a serious mismatch between how technologists conceptualise fairness, compared to other humans. How can AI regulation bridge this gap?

Traditional models of regulation cannot solve this problem. Fairness is too nuanced, too contextual, and is ultimately a human emotional response. Instead the book proposes to place the responsibility on the AI community to explain and justify their efforts to achieve fairness, basing regulatory and legal responses on how well that explanation deals with the risks that particular AI presents, and whether the AI operates in accordance with the explanation in use.

The book concludes by examining how far this regulatory model might be useful for some of the other social problems which AI generates.

An original and significant contribution to the literature on AI regulation, this book is a must-read for those working in the areas of law, regulation, and technology.

AI Fairness and Beyond

Law, Regulation, and Technology

Chris Reed

•HART•
OXFORD • LONDON • NEW YORK • NEW DELHI • SYDNEY

HART PUBLISHING

Bloomsbury Publishing Plc

Kemp House, Chawley Park, Cumnor Hill, Oxford, OX2 9PH, UK

1385 Broadway, New York, NY 10018, USA

29 Earlsfort Terrace, Dublin 2, Ireland

First published in Great Britain 2024

A catalogue record for this book is available from the British Library.

A catalogue record for this book is available from the Library of Congress.

Library of Congress Control Number: 2024934264

ISBN: HB: 978-1-50997-680-5
ePDF: 978-1-50997-681-2
ePub: 978-1-50997-682-9

Typeset by Compuscript Ltd, Shannon

PREFACE

For over 30 years my writings have focused on the regulation of computing technology, and my interest in artificial intelligence (AI) goes back even further than that. Throughout, I have consistently argued that new technology regulation is usually defective if it tries to predict the risks which that technology will create, and thus that it is better to wait until the technology has been in use long enough to identify what those risks really are.

This book is a departure from that position, which rather surprises me. I began the project simply as an investigation into whether it might ever be possible to understand, and thus regulate for, society's desire that AI should replicate some of the values which we humans have consistently been unable to explain in any detail. I describe these as aspirational values because, although we collectively understand the outcomes we would like AI to achieve, we have little idea how we might tell AI developers the ways in which we expect them to achieve those outcomes.

Fairness was an obvious choice for several reasons: it is a core value in law and regulation; it has been investigated extensively in other disciplines; and yet, no lawmaker or judge has ever, so far as I could discover, come up with a convincing definition which incorporates all the constituent elements of fairness. I thought that I might be able to develop a deeper understanding of the problem and perhaps identify how we might know when it was time to begin to regulate, and some of the regulatory choices which might be available.

I soon realised, though, that AI is developing so fast, and becoming so rapidly embedded in human activities, that it might depart too far from those human values before we realise that this had happened. If so, correction through regulation would be painful, if it were even possible. I therefore think we need to begin regulation now, even though we do not fully understand the values which regulation should aim to achieve or the ways to achieve them. That regulation needs to be based on as deep an understanding of current problems as can be achieved, rather than on speculation about what future problems might conceivably arise, and needs to take account of what AI technologists are currently able to achieve and what they cannot do. The main purpose of regulation at this stage should be to guide AI technology development in a socially desirable direction – we cannot hope to produce a definitive scheme of regulation in our current state of ignorance.

Thus the regulatory scheme which I put forward in chapter eight should be seen as merely the first step on that path. Although I believe that it would go some way towards solving the regulatory problem for the immediate future, I am also sure that it is neither a complete solution nor the final regulatory answer.

In the course of developing that scheme I realised that it might be modified to deal with some of the other aspirations which society has for AI, and so the book concludes in chapter nine by analysing how it might be used for a small number of examples.

The ideas I have expressed here have been developed over many years through discussion with academic and practitioner colleagues, often on topics with no apparent connection to AI. Those colleagues are far too numerous to list here, but I am grateful to all of them. I need to express particular thanks to Dr Keri Grieman, whose ideas on the regulation of AI which might cause death or injury (soon to be published by Hart as *Law, Death, and Robots: The Regulation of Artificial Intelligence in High-Risk Civil Applications*) helped me to form my regulatory proposal in chapter eight and to Catherine Maclean who very kindly read chapter three and reassured me that the conclusions I was drawing from the work of sociologists and psychologists were not implausible to one trained in those fields.

CONTENTS

1

Introduction

A desire, perhaps even a need, for fairness seems to be innate in the human animal. Few if any readers will not, at some time in their childhood, have cried aloud 'It's not fair!'

The origins of fairness are thought to be the result of evolution. Among humans, and almost certainly our predecessor species too, certain individuals shared surplus resources, such as food and shelter. These individuals would be more likely to be recipients of sharing by others. As a consequence of access to more resources, those who shared would be more likely to survive and leave descendants. To the extent that their predisposition for sharing, reciprocity, or both, had a genetic component, those descendants would similarly be likely to engage in sharing. Thus, because sharing and reciprocity of sharing were survival traits, they spread through the population.[1]

As human(oid) groups grew in size, sharing would have become more complicated.[2] Powerful or opportunistic individuals might take all or a larger share of the resources, putting the survival of other group members at risk. If the group's continued existence enhanced the survival and reproduction chances of its individual members, then cultural mechanisms to ensure that all members received an adequate share would also evolve.[3] Thus in modern human groups we find social norms about sharing on a basis other than power or luck – in other words, fairness norms.[4] These norms demand fairness to all members of the group,

[1] Bowles, Samuel, and Herbert Gintis, *A Cooperative Species* (Princeton NJ, Princeton University Press, 2011); for an easily accessible account see Matt Ridley, *The origins of virtue* (London, Penguin, 1997). See also Ken Binmore, *Natural Justice* (Oxford, Oxford University Press, 2005).

[2] See Binmore (n 1) 27: 'fairness evolved for use in situations in which face-to-face bargaining isn't an option'.

[3] Paul Seabright, 'The evolution of fairness norms: an essay on Ken Binmore's Natural Justice' (2006) 5 *Politics, Philosophy & Economics* 33, 40 suggests that past reciprocity may have influenced the development of these norms:

> First, by offering to share with someone else the food I have found, I may stimulate a reciprocity reaction that benefits me, whereas I would not do so if I knew that other people's behavior toward me in the future would be determined only by forward-looking considerations. Second, when some people are genuinely motivated by reciprocity, others may behave as though they were so motivated in order to gain a reputation for trustworthiness. It may take only a very small proportion of the genuinely motivated in the population to create reputational incentives for all or most of the others.

[4] Recent research suggests that bonobos, our closest animal relatives along with chimpanzees, have also evolved strong fairness norms – Jonas Verspeek and Jeroen MG Stevens, 'Behavioral and physiological response to inequity in bonobos (Pan paniscus)' (2023) 85 *American Journal of Primatology* e23455.

so that one individual will punish another for unfairness to a third party even though that unfairness had no effect on the punisher.[5]

In a small group, norms for fair sharing can be continually negotiated and agreed as part of the social life of the group. They can be varied easily to fit the context of the sharing – for example, in hunter-gatherer groups there are different norms for sharing small game such as birds, as opposed to large game like antelope. Subtle and changing variations are accommodated via the continuous dialogue between group members.[6]

Once the group reaches a certain size, however, the time and other costs of norm negotiation become too great. The group needs a decision-maker to allocate shares, and the norms (and as the group grows bigger, rules) which that decision-maker should use need to become more static and thus predictable. The larger the society, the greater the level of codification which is needed to allocate sharing. Large societies eventually develop institutions to structure decision-making.[7]

This leads us to modern societies, which from a legal perspective are usually conceptualised as the nation state. These societies have a much wider range of resources to allocate, including intangibles such as opportunity and autonomy, and require formalised rules which decision-makers must follow when making those allocations. Some examples, which will be referred to again as the book progresses, include:

- Decisions about which job applicant to employ – the job is a scarce opportunity resource because more people want it than can be employed. Here there is usually no legal duty to act fairly in deciding between applicants, but in many jurisdictions there is a legal duty not to decide on the basis of factors which are *unfair* – a duty not to engage in unlawful discrimination on such grounds as race, sex or sexual orientation.
- Decisions by financial services companies about whether to grant a loan, the resource here being money. These decision-makers are often under a positive legal duty to make fair decisions.[8]

[5] Ernst Fehr and Urs Fischbacher, 'Third-party punishment and social norms' (2004) 25 *Evolution and Human Behavior* 63.

[6] Economic experiments support that evolved conceptions of fairness manifest in these ways – see eg Herbert Gintis, 'Behavioral Ethics Meets Natural Justice' (2006) 5 *Politics, Philosophy and Economics* 5; Joel Sobell, 'Interdependent preferences and reciprocity' (2005) 434 *Journal of economic literature* 392; Samuel Bowles and Herbert Gintis, 'The evolution of strong reciprocity: cooperation in heterogeneous populations' (2004) 65 *Theoretical Population Biology* 17.

[7] These institutions might be conceptual, such as property rights, or organisational such as courts and tribunals. Frey and Bohnet have produced experimental evidence that the degree of interaction between participants in sharing decisions increases the role of fairness, and that these institutions shape those interactions and thus also shape conceptions of fairness – Bruno Frey and Iris Bohnet, 'Institutions affect fairness: Experimental investigations' (1995) *Journal of Institutional and Theoretical Economics (JITE)/Zeitschrift für die gesamte Staatswissenschaft* 286, 291–92 and 300.

[8] See the discussion in ch 5.3.3.

- Decisions about how personal data should be used by its controller.[9] The exercise by the data controller of its autonomy to use the data reduces the autonomy of the data subject to limit its use, and vice versa. Autonomy here is a scarce resource which needs to be shared, and the data controller has a legal duty to use the data, and so share autonomy, fairly.

Thus fairness has become an important element of law and regulation. If the decision-maker is a human, the law need say little more than that he or she must act fairly. Because fairness is an innate human trait whose content is determined by the society in which the decision-maker is operating (I will say more about this in chapter three), the law assumes that the decision-maker understands what fairness means. If we have doubts about the fairness of a decision, we can compare it with similar decisions by other human decision-makers in order to identify whether it was in fact fair.

If, though, we substitute an AI system for the human decision-maker and still desire fairness, we face two problems:

- the AI has no innate and evolved understanding of fairness, and thus
- we will need to specify what the AI's developers are required to do to ensure, or at least provide us with sufficient assurance, that the AI will make adequately fair decisions.

This book explores those problems in more depth, and makes some first suggestions about how they might be solved.

The starting point for the book is the innate human desire for fair decision-making. This is so important that it is often specifically demanded by law and regulation for a particular activity. As AI takes on decision-making, we may need to make the same fairness demands of it, and maybe extend the range of legal and regulatory fairness obligations. However, doing so is highly problematic. Fairness is never properly defined by law and regulation – in a human decision system it is assumed that other humans can recognise unfairness without a definition, and can assess retrospectively (though not in advance) whether a decision was fair. This presents a major problem when demanding that AI makes fair decisions, because it give no guidance to AI developers as to how to comply with regulation, or how to assess whether an AI decision system is adequately fair.

Thus we need try to understand fairness in more depth, and so I begin by investigating the social understanding of fairness, drawing mainly on the work of sociologists and psychologists. This is the basis of the standard of fairness which law and regulation hope to achieve. We will see that fairness is an emotional

[9] Defined in law as the person who makes decisions about use of personal data, EU General Data Protection Regulation (GDPR), Regulation (EU) 2016/679 of the European Parliament and of the Council of 27 April 2016 on the protection of natural persons with regard to the processing of personal data and on the free movement of such data, and repealing Directive 95/46/EC (General Data Protection Regulation), OJ L119/1, 4 May 2016, Art 4(7).

reaction to a decision, influenced mainly by a desire that like cases should be treated alike. However, merit and disadvantage both justify preferential treatment, and fairness reactions are also influenced by both power and status. How these factors combine to generate the emotion is heavily mediated by the context of the decision. All this explains why fairness cannot adequately be defined, even if unfairness can be recognised after the event.

The book then turns to how human theorists have attempted to explain fairness, by constructing simplified philosophical or economic models of the human understanding. These are abstractions of the social conception of fairness, and are potentially usable by AI developers as guides in their search for AI fairness techniques. However, there are significant differences between these theoretical explanations of fairness. They embody different understandings of fairness which are not compatible with each other. None is therefore a definitive guide for AI fairness, and choices must be made.

Our next step is to investigate some of the ways in which fairness is demanded by law and regulation. The lawmaker's aim is to ensure that decisions are at least adequate to satisfy society's conception of fairness, but law and regulation go about this in quite a restricted way. We will discover that compliance with these requirements depends very much on interest balancing, on procedures which in particular eliminate impermissible bias, and on a sufficient causal connection between the factors which determine the decision and its outcome. All these will need to be factored in to AI development.

At this point the hope is that the reader will have as complete a picture of fairness as is achievable in a book of this size, and I go on to construct a composite model of the human understanding of fairness. This represents the ideal which AI decision systems should be asked to meet, so far as possible. The complex and nuanced nature of fairness presents difficult challenges for AI development, and these are analysed and explained.

Finally, before attempting to construct a regulatory system for fair AI decision-making, we need to understand the limitations which AI development faces in achieving decision systems which correspond to the human understanding. In chapter seven I review the various approaches which the AI community has taken to this question, identifying their contradictions and shortcomings. This will help us decide what can reasonably be asked of AI producers and what cannot, but also builds a picture of how far AI fairness might differ from the conception used in human-to-human interactions whilst still being acceptable to society.[10]

[10] See Peter-Paul Verbeek, 'Some misunderstandings about the moral significance of technology' ch 5 in Peter Kroes and Peter-Paul Verbeek (eds), *The moral status of technical artefacts* (Dordrecht, Springer Netherlands, 2013) 76: 'Human beings and technological artifacts have become so closely connected in our everyday lives, that even our moral perceptions and decisions have become technologically mediated. Only by recognizing this interweaving of humans and technologies can we take responsibility or the ways in which technologies have an impact on society and on human existence – in practices of technology design, implementation, and use.'

The proposal for a regulatory system which follows is based on this broad and deep analysis of fairness. It rejects any attempt to set out rules about how AI decision systems should be designed and operate, because rule-following alone is incapable of achieving fair decisions. Instead the proposal focuses on choices made by the humans who are responsible for those decision systems. They are, quite simply, required to identify and explain the fairness risks which are created by their AI, how they have attempted to mitigate them, to justify those fairness choices, and to ensure that the AI operates in accordance with that explanation. By doing so they provide advance assurance to society and to regulators that their AI system is likely to produce fair outcomes. Any decision-making system will inevitably produce at least some unfair decision outcomes, and the quality of explanation and justification, together with the extent to which the AI actually operates as described, offers a standard against which to measure the adequacy of compliance. A range of liability and regulatory consequences for failure is proposed, together with thoughts about where regulation should apply and how different regulators might coordinate.

The book concludes by exploring how far this regulatory technique might be used for other AI risks where it is only possible to specify the desired outcome, rather than the ways in which AI producers should behave. This last chapter investigates consistency of decision-making, trustworthiness and explainability. It also suggests that regulating in this way might go some way towards dealing with social conflicts which are, at least in part, within the scope of existing law but where that law fails to address them and is difficult to change. The test case here is the use of third-party intellectual property in generative AI development and training.

Before we start it is worth addressing two issues which might otherwise confuse some readers. The first is that of bias. Readers who come to this book from a technical background might be familiar with the large body of work on AI fairness which conceptualises fairness in terms of avoiding bias. As will become apparent from the chapters leading up to the discussion of AI fairness in chapter seven, bias is only one example of unfair decision-making. The human conception of fairness is far more complex, and one of the questions which this books hopes to answer is how we can regulate for fair AI in the more nuanced way which is necessary to meet society's expectations.

The second, which might be relevant for lawyers and philosophers, is the potential to confuse justice and fairness. Philosophers tend to use the term justice to describe what lawyers see as fairness, and to use equity where a lawyer would talk of justice. This book considers fairness alone, though legal readers will see some justice/equity elements in chapter five. Wilson[11] explains how to conceptualise the difference for a human decision-maker: justice focuses on whether the decision-maker undertook their work in an appropriate manner, while

[11] Bart J Wilson, 'Contra private fairness' (2012) 71 *American Journal of Economics and Sociology* 407, 415.

fairness represents the reaction of external observers to the outcome which the decision-maker pronounced. Of course, the two shade into each other to an extent, so that in particular some failures of procedure on the part of the decision-maker can lead external observers to consider the decision was made unfairly (see section 5.1). The main point, though, is that the regulatory aim for fairness must be for AI decisions to achieve acceptance as fair by external observers. How the decision was made influences that acceptance, but regulation which focuses only on controlling or specifying the decision-making process is missing the main point, which is societal acceptance of the decision outcome.

A (Long) Note on Terminology

It is less than ideal for the reader of a book to be unsure exactly what the author is talking about. AI is a particularly difficult topic in this respect, because each reader already has some conception of what AI means, but everyone's conceptions are likely to be different. The book is about who might be made legally responsible for ensuring that AIs make fair decisions, and so I also need to explain who I am referring to in those discussions.

AI

The term 'artificial intelligence' is a contentious one.[12] What I am examining here are computer systems which are used to make decisions which were formerly made by humans, and where the decision-making is not so deterministic that the human developer of the system can explain in advance all the decisions which might be made and the circumstances in which the system would make them. These might better be described as 'autonomous systems',[13] but AI is the label commonly used both in general discussion and in the technical and academic literature. Working definitions in policy documents are deliberately vague – for example, 'An AI system is a machine-based system that can, for a given set of human-defined objectives, make predictions, recommendations, or decisions influencing real or virtual environments'.[14]

[12] See International Association of Privacy Professionals, *International Definitions of Artificial Intelligence* (2023), iapp.org/media/pdf/resourcecenter/internationaldefinitionsofai.pdf. It is noteworthy that even within a single jurisdiction, different definitions of AI have been developed for different regulatory and policy areas. Definitions tend to include concepts such as the autonomy of the technology and its replication of human-like decision-making, but there is no consistent factor linking all, or even most, of these definitions.

[13] This is the term used in the UKRI Trustworthy Autonomous Systems Programme, www.ukri.org/news/new-trustworthy-autonomous-systems-projects-launched.

[14] United Nations Conference on Trade and Development, *Information Economy Report* (2017), unctad.org/system/files/official-document/ier2017_en.pdf.

To clarify the distinction between deterministic autonomous systems and those I shall term AI, we might consider automated braking system (ABS) technology on motor vehicles. This uses computing technology to determine how a vehicle's brakes should be applied to maintain control in slippery conditions. ABS overrides the human's decisions about braking, but is not AI in the sense discussed here because all the ways in which the system will work can be explained in advance.

By contrast, the AlphaGo game-playing AI makes the kinds of decisions which humans make, and the reasons why it makes them cannot be specified in advance. AlphaGo is probably the world's most proficient player of the board game Go, and the 'reasoning' and strategy behind some of its decisions is still being debated by experts in the game.[15]

The types of AI discussed in this book are mainly what is called Narrow AI. They have been developed for, and operate only in, a highly constrained domain. For example, a recruitment AI which filters applications to produce a shortlist is not capable of any other kind of decision. Narrow AIs are already being put to use outside the research environment, and are thus the most immediate regulatory concern.

Readers should not be seduced into thinking in terms of Artificial General Intelligence (AGI). An AGI is an AI which is capable of working as a decision-maker in multiple domains, and is able to learn new domains without further human intervention. When I was trying to build simple expert systems in the early 1980s, AGI was reckoned to be around 10–20 years from successful construction.[16] In May 2023 the founder of DeepMind, Demis Hassabis, reduced this estimate to 'a few years, maybe within a decade away'.[17] Treating Hassabis's prediction with understandable scepticism, I suggest that if AGI is still 10–20 years distant it is definitely too early to try to decide how to regulate it for fairness, or indeed for any other desired characteristic. AGI will not magically appear all in one go, if it is ever achieved. Rather, it will evolve from previous AI research, and as it does so the issues which society needs to regulate will become apparent, as will the possible regulatory solutions.

Thus the current regulatory focus is best placed on narrow AI, and this is what the book discusses. Inevitably, narrow AI will evolve an increasingly broad range of uses within its narrow domain as we have seen with generative AI. Generative AI hit the public stage as this book was being written with the launch of ChatGPT as a free-to-use trial of the technology. Although ChatGPT and its rivals can look very like a form of AGI to observers who are unaware how they work, what they are doing is still very narrow in scope, though broad in application.

[15] www.deepmind.com/research/highlighted-research/alphago.

[16] See eg Ehud Y Shapiro, 'The fifth generation project – a trip report' (1983) 26 *Communications of the ACM* 637.

[17] 'DeepMind boss says human-level AI is just a few years away' *The Independent*, 4 May 2023.

Artificial Superintelligence (ASI, sometimes termed 'the Singularity') would be an AI which exceeds human capabilities in every respect. It is certainly too early to worry about regulating ASI – if it ever arrives, explaining fairness will be one of the least of the problems facing law and regulation.

AI Decisions

Throughout the book I will be referring to AI decisions, the output of AI systems which replace decisions made by a human. These are 'human out of the loop' systems – the AI's decision is implemented directly. It turns the wheel of a vehicle, grants loans, rejects employment applicants, and so on. Whenever society expects the human who has been replaced to make the decision fairly, there is an argument for considering whether the AI should also be required to be fair.

At first sight there is no need for fairness regulation for AI 'human in the loop' systems. These are where the AI's decision is either a recommendation to a human decision-maker, or is reviewed by a human who has the power to prevent it being implemented automatically. Why can we not simply rely on that human to ensure fairness? The answer depends on whether the human is realistically capable of identifying unfairness. For example, suppose that a hospital implements an AI which helps to allocate scarce medical treatment to patients, based on the patient's likelihood to benefit from the treatment. The AI gives each patient a score and thus ranks them in order of appropriateness to be treated. However, the hospital makes the final decision via a committee of senior medics who review the AI's recommendations as part of their decision-making. If the AI scores some patients unfairly (we will see later in this book a number of ways in which this might occur), that unfairness is likely to be invisible to the committee and thus not corrected in their deliberations. Thus even though the AI is only making a recommendation to humans, not a decision which is implemented directly, I suggest we might still want its recommendations to be fair and, possibly, regulate to achieve that.

Developer, User and Producer

If we regulate to achieve AI fairness, we need to regulate *someone*. The AI itself cannot hear our regulatory demands, nor can it respond to them. So we need to identify the persons who are responsible for the AI's decisions and allocate responsibility for achieving AI fairness between them. There are two groups of such persons: the *AI developer*, who designs, codes, trains and tests the AI; and the *AI user*, who puts the AI into use to make decisions. In the medical treatment AI example above, the hospital is clearly the AI user. It is likely that an independent technology provider acted as AI developer, either developing the AI specifically for the hospital or developing a more general purpose medical treatment allocation AI and then, perhaps, customising it for that particular hospital and treatment.

Of course, the demarcation between developer and user is not as clear-cut as this suggests. Unless the developer has created a pure 'off-the-shelf' AI, AI users may play a part in the development too. For a custom-developed AI, users will participate in the design, probably provide training data, and have a substantial role in testing. Even if the AI requires no modification for the user, the AI developer will in many cases assist the AI user to integrate the AI into its operations, provide training and so on, and so will share some user responsibilities.

Thus it is generally impossible to say in advance which should be responsible for a particular aspect of achieving fairness, because that depends on how the AI was created and implemented in any specific case. When I refer to AI developer I mean whichever of them (or both) is responsible for the aspect of development at issue, and the same applies when I refer to AI user.

To simplify matters where either might be responsible, I use the term *AI producer*, which means whichever of the developer or user ought to be responsible, or both if that is the case.

2

Do We Need Fair AI?

It is worth asking this question at the outset because it is possible to envisage some AI applications which, at first sight at least, seem to be making decisions on a purely objective basis. Do we really need to worry about fairness in those cases?

As a test example, let us look again at the hypothetical medical AI which assesses whether a patient is likely to respond to a particular treatment.[18] The AI has been trained on a large and representative sample of the patient population, and testing shows that its predictions are highly accurate, more so than even the best specialist medics. The treatment is expensive, and so can only be afforded for a minority of patients. On an objective basis it seems obvious that the treatment should be provided to those who are predicted to be most likely to benefit, and that fairness questions do not arise.

The obvious is not always correct, though. Suppose further that likelihood of responding to the treatment closely correlates with childhood poverty – the children of poor parents are less likely to benefit; those of rich parents much more likely – and that childhood poverty correlates closely with current wealth. The consequence of using the AI will be to allocate most of the treatment to wealthy patients. This is objectively justifiable, as they are most likely to benefit, but is it fair? Many would think not.[19]

In this case we can see that the possible unfairness is found in our reactions to the *collective outcome* of the AI's decisions. Each individual decision might be perceived as fair, but the totality of decisions, viewed as a whole, are still seen as unfair. To correct this once the problem was recognised, a social debate would be needed to agree what the fair collective outcome should be, and then the

[18] This is not an entirely fictional example – see Ziad Obermeyer, Brian Powers, Christine Vogeli and Sendhil Mullainathan, 'Dissecting Racial Bias in an Algorithm Used to Manage the Health of Populations' (2019) 366 *Science* 447.

[19] This example also illustrates that fairness is not cost-free. If society could decide a fairer but different way of allocating these treatments, based on factors other than the likelihood of benefitting medically, the inevitable outcome would be a lower benefit overall because fewer of those treated would respond to the treatment.

Further, identifying the social consensus is not a simple matter. For example, when US residents were asked whether the budget for medical screening should be spent on a more effective test for half the group of Medicaid recipients, or on a less effective test for the entire group, there was a clear consensus in favour of the latter. However, when the researchers reframed the question, the social preference changed to the former. See Peter A Ubel, Jonathan Baron and David A Asch, 'Preference for equity as a framing effect' (2001) 21 *Medical Decision Making* 180.

AI's workings would have to be modified to achieve that outcome. The unfairness was an unintended result of the design and development choices made by the AI producer, but it is hard to say that the unfairness was built into the AI itself. The AI did exactly what it was intended to do, predicting responses to treatment, and it was the human decision to allocate treatment on that basis which caused the unfairness.

However, unfairness can be built into an AI unintentionally, thus inevitably leading to a proportion of unfair outcomes. If we return to the example in chapter one of an AI which filters job applications to determine a shortlist for interview, we need to ask the fairness questions much earlier. Assume that the AI is developed via machine learning and trained on a dataset of previous applicants which labels them according to whether they were rejected initially, shortlisted or employed. The decisions about shortlisting and employment will have been made by humans, and in this field humans are notorious for preferring candidates who are similar to existing, successful employees. Thus if in the past the human decision-makers have (perhaps unconsciously) preferred some educational qualifications over others, or some racial groups over others, those preferences will be embedded in the training data. The AI's learning will merely teach it the same preferences which the human decision-makers exhibited, and if they were unfair the AI will make similarly unfair decisions.[20]

So it seems that we need to consider fairness for potentially *all* AIs, though we will decide after investigation that some do not raise fairness issues. This means that we need to define, or at least explain, fairness in sufficient detail for AI developers. Immediately, we hit a major snag.

Fairness is not an objective concept – rather it is a feeling or an emotion.[21] Individual conceptions of fairness may differ for the same situation, and so any group's fairness norms will reflect a collective view within the group,[22] rather than necessarily expressing a clear consensus about fairness in the wider population.

To complicate matters further, fairness is contextual. As an illustration, if we return to the suggested origins of the human need for fairness in hunter-gatherer communities, in times of plenty it might be fair to give the largest share of meat

[20] This is not a theoretical example – see Reuters, 'Amazon scraps secret AI recruiting tool that showed bias against women' (11 October 2018), www.reuters.com/article/us-amazon-com-jobs-automation-insight/amazon-scraps-secret-ai-recruiting-tool-that-showed-bias-against-women-idUSKCN1MK08G; Miranda Bogen, 'All the Ways Hiring Algorithms Can Introduce Bias', *Harvard Business Review* (6 May 2019) hbr.org/2019/05/all-the-ways-hiring-algorithms-can-introduce-bias.

[21] Jerald Greenberg, 'A taxonomy of organizational justice theories' (1987) 12 *Academy of Management Review* 9, 11.

[22] Experiments identify that group members do not differ in their understandings of what fairness consists of, based on their propensity for feeling emotions (affect intensity). Rather, the emotional understanding of fairness seems to be shared within that group. Van den Bos, Kees, Marjolein Maas, Ismintha E Waldring, and Gün R Semin, 'Toward understanding the psychology of reactions to perceived fairness: The role of affect intensity' (2003) 16 *Social Justice Research* 151, 164 finds that 'perceptions of fairness and justice are not influenced by individual differences in affect intensity, but that subsequent affective reactions to the perception of fairness or unfairness are strongly influenced by affect intensity.'

to those who expended most effort in hunting it. In times of famine, however, it might be more fair to give the bigger share to those who are weakest and thus in greatest need.

Both law and AI developers therefore need to work with their society's *collective* conception of fairness. Different societies understand fairness differently, as will be explained in chapter three, but each lawmaker has its own legitimating community which consists of that state's citizens and perhaps outsiders from whom obedience to that law is claimed.[23] This community is the society whose collective concept of fairness should guide lawmakers.

This does not mean that legal requirements for fairness need to match the social conception exactly. One role for law is to set the normative framework for a society, establishing standards for expected behaviour rather than modelling how the majority actually behave. Within the author's lifetime the UK has transformed from treating homosexual activity as a criminal offence to the celebration of gay marriage, and at all times in that evolution the law led public opinion. But laws which try to change social norms too quickly are opposed, ignored and subverted,[24] so maintaining a high degree of concordance between legal and social norms is important.[25]

This tells us that any general definition or description of fairness in the law needs to be close enough to the social conception within that lawmaker's community for it to be accepted by community members (this is explored in more detail in chapter three). In some areas, law has been able to provide some specificity by defining matters which require extra care and attention to achieve fairness.

The first of these is the fundamental rights of individuals (and in some cases companies), such as free speech and privacy. Fundamental rights are set out in constitutions or legislation, and/or established gradually through court decisions. They are rights against the individual's state – the state has an obligation not to diminish those rights, through laws or other state action, and an obligation to secure those rights for individuals (eg by prohibiting certain behaviours by other individuals). This has two implications for AI fairness. First, AI decisions which diminish the fundamental rights of individuals are likely, though not certain, to be unfair (see further chapter five). The uncertainty arises from the second implication, which is that because fundamental rights are often in conflict with each other[26] and with the broader interests of society,[27] choices have to be

[23] See further Chris Reed and Andrew Murray, *Rethinking the Jurisprudence of Cyberspace* (Cheltenham, Edward Elgar, 2018) 18–24.

[24] Christopher Jensen, 'The More Things Change, the More They Stay the Same: Copyright, Digital Technology, and Social Norms' (2003) 56 *Stanford Law Review* 531, 562.

[25] Dorothea Kubler, 'On the Regulation of Social Norms' (2001) 17 *Journal of Law Economics and Organization* 449, 452–53.

[26] Privacy and free speech often clash in an obvious manner, as do many other rights in more subtle ways.

[27] For example, Art 23 of the Universal Declaration of Human Rights sets out a right to work and free choice of employment, but convicted prisoners are likely to be deprived of that right. Article 12

made when balancing the relative impact on those rights and interests when making a decision which has to be fair.

Second, the law in most countries usually forbids discrimination on grounds of protected characteristics. In the UK these are set out in section 4 of the Equality Act 2010: age, disability, gender reassignment, marriage and civil partnership, pregnancy and maternity, race, religion or belief, sex, and sexual orientation. Discrimination occurs when an individual with one or more of these characteristics is, because of that fact, treated less favourably than another without the characteristic.[28] The law therefore divides the population into groups (this is important for AI development, as we will see in chapter seven) and requires equal treatment across these groups. Equal treatment, as chapters three, four and seven will explain, is one of the main issues which lie at the heart of the human conception of fairness.

A further factor which needs to be taken into account when determining fairness is the broader interests of society. In our medical AI example above, it initially seems unfair that it is society's richest members who will benefit from the treatment. However, society has an interest in obtaining the greatest collective benefit from its investment in healthcare, and arguably the AI's decisions achieve that aim. An assessment of whether the AI is acting fairly needs to consider both the unequal treatment of social groups and the need to accommodate society's interests – AI development which focused on just one of these would fail to achieve fair outcomes.

Laws which aim to achieve societal advantages will often only be able to do so by diminishing individuals' fundamental rights or treating them differently from members of different groups. A current example from the UK is the Online Safety Act 2023. The Act aims to prevent the dissemination of unlawful content and to reduce the risk that harmful content is disseminated to children. This requires monitoring and moderation of content, which inevitably diminishes the fundamental rights of free speech and privacy. Monitoring and moderation may also treat proponents of particular viewpoints in ways which would otherwise be discriminatory. All these raise fairness issues.

Detailed discussion of how law and regulation incorporate fairness requirements is left to chapter five, and at this stage it is merely necessary to note that legal analysis begins by asking if there has been a diminishment of fundamental rights, or discrimination or bias. Any of these are likely to amount to unfairness if there is no justification for them. Specific justification may be found in laws whose aim is to protect the broader interests of society. In the absence of any such laws, the questions which law asks centre around society's collective view of fairness, which is therefore our next issue to explore, and also around philosophical and economic conceptions of fairness, which will be examined in chapter four.

of the International Covenant on Economic, Social and Cultural Rights sets out a right to health, but the medical treatment AI example discussed above cannot comply in its decisions, nor can any fairness modifications to it, because the funds society allocates are insufficient for all to have the treatment.

[28] Equality Act 2010, s 13(1).

3

Society's Conception of Fairness

Psychologists investigate how individuals understand fairness, and sociologists study the collective understanding of fairness in society which derives from that set of individual understandings. Together, they can help us understand the base level of fairness which law and regulation should be seeking to achieve from AI.

The findings of researchers in these disciplines tell us that fairness in decision-making is understood at two levels: *outcome fairness* assesses whether the outcome of the decision is fair, particularly in comparison to other decisions, while *procedural fairness* assesses whether the way in which the decision was reached was fair. In theory outcome fairness is paramount – an unfair outcome is still unfair, no matter how fair the process. However, in practice, procedural fairness is often more visible than outcome fairness, and may thus play a greater role in the assessment of a decision.[29] This only applies, though, if the decision-maker's authority is accepted by the group as legitimate – if not, the focus is instead on decision outcomes.[30]

Overall, it is important to remember that outcome fairness is what society is expecting. Procedural fairness is a mechanism for achieving outcome fairness, or at least for demonstrating that outcome fairness is likely to be achieved.

3.1. Outcome Fairness

If asked to explain fair decision-making outcomes, most individuals would begin by saying something like, 'I can't define fairness, but I know unfairness when I see it.' If pressed, they might come up with concepts like equality of treatment, decisions based on merit, and decisions modified to account for disadvantage or need. These are exactly the concepts which researchers have identified more precisely as the primary constituents of outcome fairness.

[29] Kees van den Bos, E Allan Lind and Henk AM Wilke, 'The psychology of procedural and distributive justice viewed from the perspective of fairness heuristic theory' in R Cropanzano (Ed), *Justice in the workplace: From theory to practice Vol 2* (Mahwah, NJ, Erlbaum, 2001) 55, 62–63. The authors note, however, that if decision subjects do know about the outcomes in other cases, outcome fairness then has a greater influence than procedural fairness.

[30] Tom Tyler, 'Procedural Justice' in Austin Sarat (Ed), *The Blackwell Companion to Law and Society* (Oxford, Blackwell, 2004) ch 23, 442–43.

Equality of treatment is more complex than it at first appears. Lee et al note that the concept of fairness:

> [I]s based on the egalitarian foundation that humans are fundamentally equal and should be treated equally. However, how equality should be measured and to what extent it is desirable have been a source of debate in both philosophical ethics from a moral standpoint, and welfare economics from a market efficiency standpoint.[31]

These economic and philosophical debates, discussed in chapter four, have greatly influenced the approach of AI researchers and developers to fairness issues (see chapter seven). However, it is unlikely that they have has any real influence on the social conception of fairness in terms of equality.

In part, this may be because the social conception of fairness is more complex and more culturally determined than philosophical and economic theories suggest. Most humans seem to accept the principle of equality in decision-making, in the sense that like cases should be treated alike. Disagreements arise because deciding whether two cases are alike depends very much on their context. For example, in the UK most employment interview panels for large organisations would choose between candidates in part on their qualifications for the position, but would not (or at least ought not to[32]) be influenced by any family relationship between an applicant and the CEO. In other cultures that family relationship might be a permissible, perhaps even socially necessary, reason for treating the candidate differently.[33]

3.1.1. Merit, Disadvantage, Power and Status

Equal distribution achieves perceived fairness in simple social situations, such as sharing a cake, but as the context becomes more complex, pure equality is increasingly seen as unfair.[34] Psychological and sociological research into fairness has identified four factors which are likely to come into play when identifying like and unlike cases: merit, disadvantage/need, power and status.

[31] Michelle Seng Ah Lee, Luciano Floridi and Jatinder Singh, 'Formalising trade-offs beyond algorithmic fairness: lessons from ethical philosophy and welfare economics' (2021) 1 *AI and Ethics* 529, 530.

[32] See eg the UK Civil Service Commission Recruitment Principles (April 2018), civilservicecommission.independent.gov.uk/wp-content/uploads/2019/03/02aRECRUITMENT-PRINCIPLES-April-2018-FINAL-.pdf, which emphasise repeatedly that selection should be solely on the basis of the published selection criteria.

[33] See eg Murray Scot Tanner and Michael J Feder, 'Family politics, elite recruitment, and succession in Post-Mao China' (1993) 30 *Australian Journal of Chinese Affairs* 89. Culture may also vary between different types of business; see Bengt Johannisson and Morten Huse, 'Recruiting outside board members in the small family business: An ideological challenge' (2000) 12 *Entrepreneurship & Regional Development* 353.

[34] We might contrast the cake division with splitting a restaurant bill where the participants' menu choices substantially differed in cost.

Equality of treatment adjusted for merit is usually described as equity.[35] The idea here is that in allocative decisions, rewards should be distributed in proportion to contribution.[36] This seems also to be the starting point for much of philosophical and economic thought about fairness.

Unequal contribution differentiates cases so that they are no longer alike, and thus unequal reward is justified. However, although this justifies differential awards, a decision may still be perceived as unfair if it departs too far from equality. Alves and Rossi,[37] surveying US attitudes to household incomes, found that very high and very low incomes are both considered unfair, irrespective of qualifications and other characteristics. This implies that equity fairness includes some concept of maxima and minima,[38] though higher incomes can still be perceived as justified as fair if the job in question requires a high level of occupational and educational achievement. The work of economic game theorists suggests that this maximum/minimum aspect of fairness is inherent to human psychology rather than a response to need (or lack of need). Experiments in unequal division where participants have neither earned nor need the reward show that grossly unequal allocations are rejected as unfair.[39]

Disadvantage or need can also differentiate cases so that they are no longer alike. Tyler suggests that this is not motivated by sympathy or empathy, but rather by some objective conception of need which would lead to injustice if that need were not accounted for in the decision.[40] Law expressly recognises some types of disadvantage or need, for example in the case of physical or mental disability where UK employers must not discriminate unjustifiably against those persons in hiring and must make reasonable adjustments to working practices and environments to account for the disability.[41]

Societal fairness judgements involve trade-offs between merit and need. As an example, Alves and Rossi[42] found that among US residents the perceived fair

[35] See Tyler (n 30) 438–39 for a summary of understandings of the concept.

[36] And also potential contribution, which justifies basing employment applications in part on qualifications and experience.

[37] Wayne M Alves and Peter H Rossi, 'Who should get what? Fairness judgments of the distribution of earnings' (1978) 84 *American journal of Sociology* 541.

[38] ibid, 557: 'In sum, respondents were both more generous and more democratic than public policy and the empirical income distribution. The ceilings on earnings represented by the computed social maxima are well below the earnings of the affluent. At the same time fair earnings for persons on the bottom of the occupational and educational pyramids are set higher than poverty-level definitions. Implicitly respondents define an equitable earnings distribution that has a higher mean and a lower variance.'

[39] Empirical experiments mainly use the ultimatum game which involves two players, one of whom has the task to divide a small pot of money between them. If the other rejects the division, neither receives any money. Pure self-interest should lead to acceptance of even the most unequal division, but experimenters have consistently found that substantially uneven divisions are consistently rejected. See eg Michel JJ Handgraaf, Eric van Dijk, Henk AM Wilke and Riël R Vermunt, 'Evaluability of outcomes in ultimatum bargaining' (2004) *Organizational Behavior and Human Decision Processes* 95.

[40] Tyler (n 30) 439.

[41] UK Equality Act 2010, ss 15 and 20.

[42] Alves and Rossi (n 37) 556–57.

minimum for one-person working households was around the median wage, but only just at the poverty line for one-person households on benefits, suggesting that the merit contribution made by working justifies a greater share than does need on its own. One suggestion about whether merit is prioritised over disadvantage or vice versa comes from Greenberg,[43] that the choice is determined by the aim of the decision-maker. Thus aiming to achieve social harmony within a group might require both merit and need to be discounted, resulting in equal allocations, whilst a sharing between friends might choose to prioritise disadvantage or need.

Power and status also affect fairness perceptions in a complex interaction. The choice by a decision-maker of the criteria used to decide fairness is itself an exercise of power, and this is equally as true of those who design fairness algorithms as it is of human decision-makers.[44] But the existence of power on the part of a decision subject can also be perceived as a justification for treating that person more favourably, and thus that it is fair to do so. It has been demonstrated that individuals who are perceived as powerful are often granted a greater share of resources by those less powerful.[45]

The effects of power can be subtle, and seem to derive from its mere existence even if it is not exercised. Lennon and Rosenfield investigated US housewives' attitudes to the division of housework[46] and found that wives who felt dependent on their husbands for resources believed it was fair for husbands to undertake a smaller share of housework. By contrast, as women's incomes rose their perception of a fair sharing moved towards equal shares. 'Women who feel that they have fewer alternatives to marriage and women whose earnings would put them below the poverty threshold if divorced are more likely to view an unequal division of housework as fair.'[47] In general, women spent more time on paid work and housework combined, but men contributed more financially. This gave men more power, and perhaps a higher status as 'breadwinner', which legitimated the unequal distribution.

Ridgeway argues[48] that the perception that power equates with merit arises through consolidating power with status. This legitimates the privileging of power because the powerful individual is perceived as a 'better' category of person

[43] Jerald Greenberg, 'A taxonomy of organizational justice theories' (1987) 12 *Academy of Management Review* 9, 13.

[44] Maximilian Kasy and Rediet Abebe, 'Fairness, equality, and power in algorithmic decision-making' (2021) *Proceedings of the 2021 ACM Conference on Fairness, Accountability, and Transparency. FAccT '21*, 576; Alycia N Carey and Xintao Wu, 'The Fairness Field Guide: Perspectives From Social And Formal Sciences' (2022) *AI and Ethics* 1, 19. This issue is examined further in some detail in Luciano Floridi, Josh Cowls, Thomas C King and Mariarosaria Taddeo, 'How to Design AI for Social Good: Seven Essential Factors' (2020) 26 *Science and Engineering Ethics* 1771.

[45] Shane R Thye, 'A status value theory of power in exchange relations' (2000) *American Sociological Review* 407.

[46] Mary Clare Lennon and Sarah Rosenfield, 'Relative Fairness and the Division of Housework: The Importance of Options' (1994) 100 *American Journal of Sociology* 506.

[47] ibid, 525.

[48] Cecilia L Ridgeway, 'Why status matters for inequality' (2014) 79 *American Sociological Review* 1.

who thus deserves more because of their merit. Power is precarious[49] but cultural status beliefs persist. Thus transforming power into high cultural status[50] confers merit on that person, which results in the perception that it is fair they should receive a greater share than those of lower status.

> Elites, for instance, signal their class status superiority through sophisticated speech, clothing, and tastes in Art … Development of cultural status beliefs about group differences, then, partially disaggregates those differences from the direct control of resources and power and gives those differences, as status distinctions, independent causal force. This, in turn, creates a reciprocal causal interdependence between cultural status beliefs about social groups and material inequalities between these groups.[51]

Conversely, where resources are shared between those of equal status, the expectation is that each will receive an equal share.[52]

The influence of status on fairness perceptions varies between cultures, so that differing attitudes to status also moderate social perceptions about the appropriate balance between merit and disadvantage when determining fairness.[53] Even within a single culture, high-status individuals tend to prioritise merit when assessing fairness while low-status individuals give more weight to disadvantage.[54] This is one of the few divides in society's collective conception of outcome fairness, which suggests that it might be appropriate to eliminate power and status considerations from AI decisions so far as this is achievable.

3.1.2. Collective Outcome Fairness

Thus far we have focused mainly on perceptions of fair decision-making in individual cases, asking whether a particular decision is seen by decision subjects or outsiders as fair. But what if the whole decision-making system is unfair at the wider, collective level? We have already seen one way in which this might happen, through individuals' perceptions that power and status are reasons for more favourable treatment. Whilst decisions to favour power and status might

[49] ibid, 3: 'inequality based purely on organizational control of resources and power is inherently unstable. It gives rise to a constant struggle between dominant and subdominant individuals. To persist, that is, for inequality to become durable inequality, control over resources and power has to be consolidated with a categorical difference between people such as race, gender, or life style.'

[50] The process through which this legitimation of status occurs is explained more fully in Joseph Berger, Cecilia L Ridgeway, M Hamit Fisek and Robert Z Norman, 'The legitimation and delegitimation of power and prestige orders' (1998) *American Sociological Review* 379.

[51] Ridgeway (n 48) 4, citing Pierre Bourdieu, *Distinction: A Social Critique of the Judgment of Taste* (trans R Nice) (London, Routledge, 1984).

[52] Shaun Nichols, 'Emotions, norms and the genealogy of fairness' (2010) 9 *Politics, Philosophy and Economics* 275, 288–90.

[53] Pamela Jakiela, 'Social preferences and fairness norms as informal institutions: Experimental evidence' (2011) 101 *American Economic Review* 509, finding that in dictator game experiments conducted in the US and in Kenya, US participants rewarded merit but placed no value on unearned status, whilst Kenyan participants were strongly influenced by unearned status.

[54] Alves and Rossi (n 37) 559.

be accepted as fair in each individual case, society will notice if overall the rich, successful and powerful are favoured consistently.

Perhaps surprisingly, perceptions of fairness in individual decision-making seem not to be shaped by political perspectives. Within any given society, researchers find a high level of consensus about individual fairness across political divides. However, the collective effects of individually fair decisions do raise political issues, because all decisions about the allocation of resources are likely to lead to inequalities from some political perspective. For example, a first-come first-served approach to medical treatment will be criticised as unfair by those whose political views prioritise merit in the sense of likelihood to benefit from it, and also but on entirely different grounds by those whose politics prioritise disadvantage.

The distinction here might be described as one between individual fairness (or justice) and collective fairness.[55] Tyler notes that: 'One of the paradoxes of justice research is that people endorse equity on the micro-level, but do not endorse the macro-level inequality that results from the micro-level operation of equity'[56] and in later writings notes that while ideological differences between individuals affect how they perceive the collective fairness of outcomes, there seems to be a consensus about procedural fairness across ideologies.[57]

This paradox is not soluble by AI developers, but is something they need very much to be aware of. The choices made by decision-makers are themselves an exercise of power,[58] by determining what balance to strike between merit and disadvantage and how to control for the effects of individual power and status. AI developers make the same choices when deciding what fairness factors to label in their training data and in how they choose to test for fair outcome distributions, but with little recognition at present that those choices have political and social outcomes which may not match those of wider society.

Ultimately it has to be society which constrains that power by deciding which choices are politically and socially acceptable, and this is normally done via law and regulation. This social debate cannot realistically be undertaken until society has experienced AI decision systems in action for some time, and so the regulatory proposals in chapter eight make no attempt to deal explicitly with the issue, though it is to some extent encompassed implicitly. It is, though, potentially an

[55] See Tom Tyler, 'Justice and effective cooperation' (2012) 25 *Social Justice Research* 355, 371–72 discussing collective action to remedy this kind of injustice.

[56] ibid, 362.

[57] Tyler (n 30) 448, citing Günter Bierbrauer, 'Political Ideology and Allocation Preferences: What do Turkish Immigrants in Germany Deserve?' Potsdam, Germany: International Network for Social Justice Research Conference (July 1997); Randall Peterson, 'The role of values in predicting fairness judgments and support of affirmative action' (1994) 50 *Journal of Social Issues* 95; Tom Tyler, 'Governing amid diversity: The effect of fair decision-making procedures on the legitimacy of government' (1994) 28 *Law and Society Review* 809.

[58] Joseph Berger, Morris Zelditch, Bo Anderson, and Bernard P Cohen, 'Structural Aspects of Distributive Justice: A Status-Value Formulation' in Joseph Berger, Morris Zelditch and Bo Anderson (eds), *Sociological Theories in Progress*, vol 2 (Boston, Houghton Mifflin, 1972) 119.

issue which fairness regulators will want to deal with in their codes and guidelines, and in due course it may require explicit regulatory action.

3.1.3. Outcome Fairness and AI Development

AI developers face a difficult challenge in their aim to achieve outcome fairness. Because society's conception of outcome fairness is so nuanced and highly context-dependent, it is impractical to assess each individual decision outcome for fairness.[59] Because the same is true for decisions made by humans, the solution sometimes adopted in human decision-making systems is to institute an appeals process. However, appeals processes recognise that issues like fairness are so dependent on their context that the decision of the human who originally made the decision is only overturned in the most egregious cases. Thus appeals tribunals tend to substitute their own decision on fairness only where they form the view that no fair and reasonable decision-maker could have reached the original decision.[60]

All this suggests that regulation which simply requires AIs to make fair decisions does not give sufficient guidance to AI developers, and thus is unlikely to assist them to make a fair AI. A more fruitful approach might be one which requires the AI developer to explain the ways in which it attempted to achieve fair decisions by the AI. This means adopting development processes which recognise the fairness issues and attempt to deal with them. Any explanation will need to reassure society that outcome fairness is likely to be achieved for the vast majority of the AI's decisions. This requires documenting and justifying the important fairness choices which the developer has made.

The first of these is the choice about which criteria are to be labelled and used to train the AI. To some extent those choices have an objective element, being defined by the domain of application. Thus a credit granting AI will focus mainly on ability to repay the loan, whilst an employment selection AI will focus on qualifications and relevant experience.[61] However, some domains do not have clearly accepted criteria to be used in decision-making. Our hypothetical AI which selects for medical treatment might be a useful example, with choices needing to be made about matters such as the level of importance to be assigned to increased life expectancy versus immediate relief from pain.[62] Although the

[59] AI developers are addressing this problem, examined in detail in ch 7, but as we will see their efforts so far capture only a fraction of the complexity of the social understanding of outcome fairness.

[60] See ch 5.2.

[61] But both qualifications and relevance of experience have a heavily subjective element – for example, should the developer equate all degrees of the same class or differentiate between the universities which awarded those degrees, and how far does management experience in a different sector equate to management experience in the same sector?

[62] Though here there is at least one consensus measure in the QALY, quality adjusted life year: One quality-adjusted life year (QALY) is equal to 1 year of life in perfect health. QALYs are calculated by

AI user is clearly entitled to set their own criteria for a positive outcome, it will be important that these should be justifiable if potential fairness is to be demonstrated.

The second is the choice about how much weight is to be given to disadvantage, and which disadvantage factors are selected for this purpose. National equality laws set out a minimum set of factors which are prohibited from forming the basis for a decision unless that can be justified,[63] and this means that the training dataset and selection criteria need to be analysed to make sure that decisions are not based on proxies for those unlawful factors.[64] Merit factors also have the potential to be proxies for unlawful discrimination factors, particularly if they are induced by the AI during the machine learning process rather than set expressly by the AI designer.[65]

Merely avoiding unlawful discrimination may not be enough, though. As we saw earlier, society's conception of fairness seems to require that the rewards for merit should not be so high that the disadvantaged fail to receive some minimum share,[66] and also that the collective set of outcomes from the AI's decisions should meet the social consensus about what a fair distribution consists of. Defining in law and regulation how AI producers should approach this issue is impossible, except perhaps for specific AI applications in already regulated sectors.[67] Ethical guidelines for AI development are likely to be useful here – they do not tell AI developers which choices to make, but can assist them to decide *how* to make those choices and justify them.

Perhaps the most difficult decision for AI producers is to decide what, if anything, to do about society's tendency to consider it fair that power and status should achieve greater allocations. If nothing is done, machine learning based on a dataset of past human decisions will inevitably produce an AI model which embeds the power and status preferences of those humans. This model will match the social conception of individual outcome fairness, though it might also produce results which are collectively considered to be unfair.

From an ethical perspective it seems obvious that power and status confer no particular merit on a decision subject, and that society's attitudes here are unwarranted. AI developers might therefore be justified in trying to control for

estimating the years of life remaining for a patient following a particular treatment or intervention and weighting each year with a quality-of-life score (on a 0 to 1 scale). It is often measured in terms of the person's ability to carry out the activities of daily life, and freedom from pain and mental disturbance. (UK National Institute for Health and Care Excellence Glossary, www.nice.org.uk/glossary?letter=q).

[63] See n 28 and accompanying text.

[64] As an example relevant to credit and insurance, an address can sometimes be a proxy for race.

[65] An example here might be athletic achievements in the case of an employment selection AI, which could indirectly discriminate against physically disabled people.

[66] See eg Ubel et al (n 19); Alves and Rossi (n 37).

[67] Our AI which allocates medical treatment might be an example here. A regulator in the medical sector could set maxima and minima for allocations between defined groups, balancing between potential benefit from treatment and proportionate distribution across groupings based on geography, income, etc. This kind of weighting exercise against defined metrics is already a core AI activity.

these factors. There are clear difficulties in doing so, particularly as objectively justifiable merit criteria such as educational qualifications also confer status and reflect power.[68] For allocations of public goods, such as access to medical treatment, any preference based on status seems hard to justify. However, in a private sector activity such as a high-end auction house, it is possible to make at least some justification for employment influenced by status because of the perceived benefit when employees share a background with the employer's clients.

Currently, controlling for power and status is not common in law and regulation though it may be addressed at a sectoral level.[69] However, ethical guidelines might usefully set out principles and good practices for AI developers here (see further section 4.1.1). One practice which might be encouraged by regulators could be the capturing and labelling of markers for power and status in the training dataset. This might enable outcomes of decisions to be monitored on a collective basis to demonstrate that, overall, the AI meets society's collective fairness conceptions.

This analysis suggests that AI developers need to work with AI users to ensure that their development procedures can satisfy society that fair outcomes are likely. Each AI is likely to need a different answer. This is because fair decision outcomes are determined by the developer's and user's choices among all the fairness considerations discussed above, taking appropriate account of the context surrounding the matter to be decided. Thus it is impossible to define a set of rules which will lead to a fair outcome in each case, nor to develop some universal standard for fair AI development which goes beyond general principles for fair development.

Although the aim has to be fair outcomes, most of the individual outcomes of an AI's decisions are likely to be hidden from outsiders. Observers will have little or no data on which to assess the overall outcome fairness of that AI. As a consequence, the most visible and important influence on social acceptance of the AI's fairness will be the justification by each AI producer of the choices made, explaining why those choices are likely to lead to fair decisions. This means that AI developed through machine learning poses particular challenges. An AI which is trained via machine learning on a sufficiently large dataset of fair decisions would be likely (though not certain) to decide fairly in most instances, because it will be modelling the social conception of fairness already embedded in its training data. However, the justification for the decisions which fairness demands

[68] Assuming intellectual ability to be fairly equally distributed across all UK groups, the fact that advantaged applicants are several times more likely to secure university admission than socially disadvantaged applicants (See Sutton Trust, *Rules of the Game: Disadvantaged students and the university admissions process* (2017) www.suttontrust.com/wp-content/uploads/2019/11/Rules-of-the-Game.pdf) indicates that access to a university degree has some close link with power and status.

[69] As a UK example, the Office for Students which regulates some aspects of universities has a policy objective to increase the proportion of socially disadvantaged people who have access to higher education. See OFS Annual Review 2019, *A new approach to fair access, participation and success*, www.officeforstudents.org.uk/annual-review-2019/a-new-approach-to-fair-access-participation-and-success.

will not be explicit, and so will need to be elicited from the AI model through technical analysis tools.[70]

We have already seen that society is more likely to judge the fairness of decision systems on the basis that they operate fair procedures,[71] to which for AI we can now add procedures for justifying the developer's fairness choices. Procedural fairness is therefore our next topic of discussion.

> In general, while people create rules to maintain relationships, no single rule can cover all situations. There are conflicts among rules. A worker with a sick child invokes the principle of need to justify leaving their job, whereas the manager invokes equity to argue that the lost productivity should be held against them. No single rule is sufficiently complex to handle the problems and issues arising in social interactions. A better approach is the creation of authorities with the ability to do what is right within the situation. Authorities have the advantage of being able to exercise discretion – that is, they can adapt to unique or changing situations by deciding to apply different justice rules or by creating hybrid approaches to justice.
>
> Hence, it is not surprising that one commonly chosen approach that groups adopt to determine 'fair' approaches to allocation is to choose an authority and then let them decide.[72]

This choosing of an authority, whether human or AI, only works if that authority is accepted as a fair decision-maker, and this is determined very much on the basis of its procedural fairness.

3.2. Procedural Fairness

We noted at the beginning of this chapter that, because decision subjects rarely have access to the outcomes of other decisions, they are likely to focus more on the procedures used by the authority which is making their decision. We need, therefore, to identify what these procedures require in order for the authority to be accepted as acting fairly.

The role which procedural fairness plays in the social domain is that it helps to achieve acceptance of decisions by their subjects.[73] The mere granting of authority to a decision-maker in a command and control model does not confer legitimacy on that decision-maker – this requires acceptance of authority by the community of those who are subject to the decisions[74] and the perceived fairness of the decision-making process plays a vital role in that acceptance.[75]

[70] See ch 9.4; Chris Reed, Keri Grieman and Joseph Early, 'Non-Asimov explanations – regulating AI through transparency' in *Law in the Era of Artificial Intelligence: Nordic Yearbook of Law & Informatics 2020–2021*, ch 14, irilaw.org/2022/02/16/new-publication-nordic-yearbook-of-law-and-informatics-2020–2021.

[71] See text to n 29.

[72] Tyler (n 55) 364 (references omitted).

[73] Tyler (n 30) 440.

[74] Reed and Murray (n 23).

[75] Tyler (n 30) 442–44.

There is an important distinction to be made here from the legal approach to procedural fairness as discussed in section 5.1. The law focuses on establishing general (and domain-specific) rules of procedure which are intended to achieve some abstract and objective concept of fairness, and largely assumes that following them will ensure a fair decision. By contrast, the work of psychologists and sociologists indicates that the focus of decision subjects (and thus society more widely) is at least as much[76] on whether the actual process undertaken convinces them that the decision was made fairly.[77] This is sometimes described as interactional fairness.

The empirical work of van den Bos et al[78] explains that a decision-subject's assessment of fairness, including procedural fairness, does not derive primarily from the application of procedural rules but is instead a heuristic activity,[79] focusing on their own experience of the decision-making process. Thus there is no procedural formula which works universally to convince society that it is fair. Rather, the manner in which each decision's procedure was implemented determines whether that procedure works to achieve a perception of fairness, and thus acceptance of the decision. How the decision subject is treated during the decision-making process seems to be at least as important as compliance with any formal procedural rules.

3.2.1. Interaction in the Decision Process

Tyler[80] notes that:

> [F]our elements of procedures are the primary factors that contribute to judgements about their fairness: opportunities for participation (voice), the neutrality of the forum, the trustworthiness of the authorities, and the degree to which people receive treatment with dignity and respect.

The mere act of participation, allowing the decision subject to make representations and to comment on the evidence, the facts and the arguments of others,

[76] There is disagreement about the relative importance of formal procedure and the manner in which that procedure is carried out. Compare the findings of Riël Vermunt, Willem Van der Kloot and J Van der Meer, 'The effect of procedural and interactional criteria on procedural fairness judgments' (1993) 6 *Social Justice Research* 183 with those in Robert Bies and Debra Shapiro, 'Interactional fairness judgments: The influence of causal accounts' (1987) 1 *Social Justice Research* 199.

[77] John W Thibaut and Laurens Walker, *Procedural justice: A psychological analysis* (Mahwah NJ, L Erlbaum Associates, 1975) laid the foundation for this understanding, and subsequent work building on that is reviewed in Tyler (n 30) 441.

[78] n 29.

[79] See Allan E Lind, Carol T Kulik, Maureen Ambrose and Maria V de Vera Park, 'Individual and corporate dispute resolution: Using procedural fairness as a decision heuristic' (1993) *Administrative Science Quarterly* 224, 226: 'The fairness-heuristic explanation suggests that people use their evaluations of process and outcome to generate a global impression of the procedural fairness of an authority, which is then used to determine whether or not the authority should be obeyed.'

[80] Tyler (n 30) 445–47. Empirical support for this proposition is found in van den Bos et al (n 29).

increases perceived fairness of the process. This is true even if the participant does not believe their participation is likely to influence the decision: 'Interestingly, procedures giving disputants a voice in the decision-making process tend to enhance the acceptance of even unfavorable decisions.'[81]

Neutrality is achieved if there is a perception that decisions are made on the basis of rules and facts, rather than on the personal prejudices of the decision-maker. Formal rules for neutrality, such as those found in the legal conception of procedural fairness (see section 5.1) of course play a part. However, far more important is the relationship established between decision-maker and decision subject, which is what mainly convinces the subject about neutrality through their experience of previous decisions, the approach and conduct of the decision-maker, and so on.

Trustworthiness is also based mainly on the interactions which take place in the decision-making process, though of course the fact that the decision-maker is trusted by a higher authority to make the decision gives some further reassurance of trustworthiness.

> People recognize that third parties typically have considerable discretion to implement formal procedures in varying ways, and they are concerned about the motivation underlying the decisions made by the authority with which they are dealing. They judge whether that person is benevolent and caring, is concerned about their situation and their concerns and needs, considers their arguments, tries to do what is right for them, and tries to be fair. All of these elements combine to shape a general assessment of the person's trustworthiness.[82]

The decision-maker's reasons for and justification of the decision play a large role here. These need to acknowledge the arguments made by parties, and explain how they have been considered and why they were accepted or rejected. Explaining the decision-maker's reasoning helps reassure the decision subject that a genuine attempt has been made to do right in a fair way.

Finally, although dignity and respect in the treatment of the decision subject has no bearing on outcome fairness, it seems to be highly important in generating acceptance of that outcome.[83] This may be because treatment with dignity and respect indicates that the subject is included in the process, rather than being a mere observer and excluded from any part in the decision. This kind of inclusion outweighs the effects of any economic disadvantage arising from the decision by a substantial margin.[84]

[81] Jerald Greenberg, 'A taxonomy of organizational justice theories' (1987) 12 *Academy of Management review* 9, 14. See also Pankaj Aggarwal and Richard P Larrick, 'When consumers care about being treated fairly: The interaction of relationship norms and fairness norms' (2012) 22 *Journal of Consumer Psychology* 114, finding that the other aspects of interactional fairness can also lead to acceptance of adverse decisions.

[82] Tyler (n 30) 446.

[83] ibid, 447.

[84] See van den Bos et al (n 29) 54.

3.2.2. Procedural Fairness and AI Development

This social conception of procedural fairness presents complex challenges for the development of fair AI. Of the four factors identified above, neutrality is perhaps the only one which an AI might confidently be expected to fulfil.

As we saw, decision subjects will be convinced about neutrality if they believe the decision will be made on the basis of rules and facts rather than personal prejudices. AIs have no personal prejudices, and society generally recognises this to be true of machine decision-making. However, an AI's training might unintentionally incorporate the prejudices of humans whose decisions are represented in the training dataset. At least in theory, AI developers should be able to convince society of the AI's neutrality by suitable reassurances that unintended biases have been eliminated. Actually achieving such elimination is no simple task, though, and this is examined further in section 7.4.

Trustworthiness is more problematic, but perhaps not excessively so. From Tyler's formulation[85] a trustworthy AI decision-maker ought to be perceived as benevolent and caring, and concerned about the subject's situation. These are human attitudes which are unlikely to be replicated in an AI, and it seems plausible that subjects will recognise this and not expect them. The AI ought also to convince them that it is trying to do right and be fair,[86] and this should be achievable through high-level explanations of the aims of the AI developer and how it attempted to meet those aims in the development process. Trustworthiness of the decision outcome can be demonstrated by giving reasons for the AI's individual decision, although this too is a complex task (see chapter seven).

The other two factors, participation and respect, derive their social acceptance from the interactions between decision-maker and decision subject. Participation here is clearly much more than merely the decision subject providing data to the AI. Procedurally fair participation requires the decision-maker to consider all those matters which the decision subject thinks relevant, even if some are then rejected for consideration. It also demands that the subject should be able to comment on and correct the other information which the AI will use as the basis for its decision. This is not currently how most AI decision-making operates, though perhaps the recent sudden leap in capability of deep learning language models, exemplified by ChatGPT,[87] might enable the construction of interfaces that allow some elements of the open-ended participation which the social conception of fairness desires. We should also recognise that AI decision systems

[85] n 82.

[86] See Allison Woodruff, Sarah E Fox, Steven Rousso-Schindler and Jeffrey Warshaw, 'A qualitative exploration of perceptions of algorithmic fairness' in (2018) *Proceedings of the 2018 CHI Conference on Human Factors in Computing Systems*, finding through qualitative research into the attitudes to algorithmic fairness of minority groups which have the potential to be treated unfairly that unfairness in algorithmic decision-making generates substantial distrust both in the algorithm and the organisation which deploys it.

[87] openai.com/blog/chatgpt.

operate across a wide range of domains, where differing levels of participation are appropriate. As one example, it is clearly undesirable that an autonomous vehicle should seek the views of its human occupants before undertaking emergency braking. As another, our hypothetical AI which allocates medical treatment is likely to be operated by a medic, not the decision subject, and thus a patient's participation would be achieved through human-to-human consultation with that medic rather than with the AI. However, if an AI were given sole decision-making power for criminal sentencing, it would clearly be essential to allow the decision subject a high level of participation and to work out ways in which that participation could be made meaningful.

Dignity and respect, in the sense discussed above, may not be achievable at the level of the AI itself. These are emotional states deriving from the process, based on a shared sense of human vulnerability. However, the wider process of which the AI is a part might be designed to take account of the need for dignity and respect, particularly in the way in which a decision outcome is communicated to the subject. 'Computer says no!', without further explanation and justification, is far from respectful.

This discussion highlights the need for justificatory explanations of an AI's decision-making, both at the general level of its aims and how they have been achieved in development, and as justifications for each particular outcome. These explanations may prove to be the main tool available to AI producers through which they might match the social conception of procedural fairness.

It also suggests that in those cases where procedural fairness is in doubt and the decision has serious consequences for its subject, some appeal process to a human would be worth considering. Appeal processes are an important element of the legal conception of fairness too, which further reinforces the case for their inclusion where appropriate.

3.3. A Social Fairness Model?

The social conception, as we have seen, is both complex and mutable. Although its basis lies in equal treatment, society's understanding of equality is modified both by merit and disadvantage. It also responds to power and status, though I have made a normative argument for why AI producers should attempt to eliminate those considerations from their AIs. An AI which produces individually fair decisions might still be perceived as unfair if its collective outcomes do not generate the emotion of fairness in observers. And, worst of all from the perspective of AI producers, social fairness perceptions change when the context surrounding a decision changes, based on unpredictable factors which are unlikely to be captured and used in the AI's decision-making. Producing an AI which satisfies the social demand for fairness seems an impossible task.

All may not be lost, though. We also saw that decision-making procedures heavily influence social fairness perceptions, helping to convince society that the decision outcomes they have not seen are still likely to have been fair. AI producers can work on enabling participation by decision subjects and can communicate decisions in a respectful way which preserves the dignity of subjects. They can attempt to convince observers of the neutrality and trustworthiness of their AIs by explaining how the development process aimed to achieve those values.

Explanations about the AI's development, particularly those which explain fairness choices and justify them, will be particularly valuable here. Those who are subject to human decision-making expect to be informed about the bases on which those decisions are made, and to have the outcomes explained to them with justification. Doing the same for AI decision systems is likely to be the best way to achieve their social acceptance.

Without more, though, AI producers are still faced with a difficult task. Their primary aim is to achieve equal treatment of decision subjects, making appropriate allowance for merit and disadvantage. There are many ways in which this might be attempted, as chapter seven will explain, all based on different assumptions and resulting in different resource allocation between decision subjects. The social conception of fairness only tells AI producers that they should select their fairness techniques so as to achieve the most favourable emotional response from society.

Can we offer more guidance on fairness choices than that? This is where the work of philosophers and economists comes into play, and we will examine their findings in the next chapter.

4

Theoretical Insights from Philosophy and Economics

The preceding analysis of the social conception of fairness is potentially helpful in deciding if an AI incorporates fair procedures, and in assessing after the event whether it produces fair outcomes. However, it does not help greatly to identify in advance the rules and principles which AI developers should adopt in order to achieve fair outcomes.

The work of philosophers and economists has therefore attracted substantial attention from the AI community, because their writings seem to set out rules and principles for fairness which are capable of being built into AI design. Philosophers derive their rules by constructing normative arguments about how fairness can be achieved, and economists develop theoretical models about how incentives and the environment affect human behaviour and test them through experiment. AI developers seek rules and models for fairness to help with AI design,[88] and thus find the work of philosophers and economists attractive for that purpose.

Philosophical understandings of fairness differ from the social and psychological descriptions we have just examined in two important ways. First, although they are grounded in the human conception of fairness, they do not describe it, but rather abstract it to construct a model of fairness based on principles which seem to correlate with how humans actually behave. This process of abstraction loses much of the detailed nuance we have seen so far. Second, there is a normative element to philosophy – it does not describe how humans actually act, but how the philosophers consider they *ought* to act in order to accord with those principles. Thus the philosophical understanding of fairness is simpler and more deterministic than the messy social and psychological understanding, and this simplification assists AI developers to devise fairness techniques which might be implementable in AI systems.

[88] Pak-Hang Wong, 'Democratizing algorithmic fairness' (2020) 33 *Philosophy & Technology* 225, 227:

> [R]esearchers and developers [have to] presume ideas of 'fairness' as a benchmark for their works, eg the definition of fairness in non-discrimination regulations, for without accepting some ideas of fairness, it is unclear what researchers and developers are programming into an algorithm and what normative standards they are using to assess whether an algorithm is fair or not.

The same process of simplification is seen to an even greater extent in the work of economists on fairness. They explicitly construct simple models which they hope might correspond with some aspect of human behaviour or of a philosophical model, and then test them through experiment, modifying their own models in consequence of the results. Thereby they hope to gain some greater insight into the drivers of human behaviour and the consequences of following philosophical models. Again, this simplification greatly assists AI developers. Even more usefully, economists can sometimes predict and test how far particular AI fairness techniques are likely to achieve their aims.

4.1. Philosophy of Fairness

4.1.1. AI Ethics

Ethicists have not ignored the problem of AI fairness, and in 2019 Mittelstadt identified at least 84 AI ethics initiatives which produced published reports.[89] These have converged on the four principles of medical ethics: respect for human autonomy; prevention of harm; fairness; and explicability.[90] It is worth noting that all these principles are normative statements about the outcomes which a fair AI *ought* to achieve. Ethics does not normally concern itself with the question of how – the assumption is that others will translate these principles top-down into AI technology, an approach which is not without its critics. John-Matthews et al point out that under AI principlism:

> [T]heoretical moral frameworks are produced deductively from abstract principles and then applied to practices. Recent studies criticize principlism for not sufficiently taking into account the particularities of these algorithms and the context of their development and warn against the side effects that can be induced when top-down measures are too prevalent.[91]

This top-down approach results in an abstract model which aims to give effect to the relevant principle when implemented in an AI. That model is inevitably defective because the process of abstraction removes the wider context which gives the principle its meaning in the world. In addition, the AI developer profession lacks the fiduciary obligations which underpin medicine, and there are no professional structures like those in medicine which both guide ethical practice

[89] Brent Mittelstadt, 'Principles alone cannot guarantee ethical AI' (2019) 1 *Nature Machine Intelligence* 501.
[90] See eg the recommendations of the High Level Expert Group on Artificial Intelligence, *Ethics Guidelines for Trustworthy AI* (European Commission, 2019), discussed below.
[91] Jean-Marie John-Mathews, Dominique Cardon and Christine Balagué, 'From reality to world. A critical perspective on AI fairness' (2022) 178 *Journal of Business Ethics* 945.

and sanction breaches.[92] Thus principles which were designed to guide behaviour within the constraints of fiduciary relationships are transformed into simplistic rules to implement, so far as possible, in code.

A useful starting point to identify core AI ethics principles is the UK House of Lords Select Committee on Artificial Intelligence report of 2018,[93] which reviews much of the previous work on AI ethics.[94] From that review the Committee suggests five overarching principles which might form the basis of an AI code:

(1) Artificial intelligence should be developed for the common good and benefit of humanity.
(2) Artificial intelligence should operate on principles of intelligibility and fairness.
(3) Artificial intelligence should not be used to diminish the data rights or privacy of individuals, families or communities.
(4) All citizens have the right to be educated to enable them to flourish mentally, emotionally and economically alongside artificial intelligence.
(5) The autonomous power to hurt, destroy or deceive human beings should never be vested in artificial intelligence.[95]

Although fairness is an element of these principles, the report does not discuss what it might mean in the AI context, or how it could be achieved.

The same is true of other, high-level statements about AI ethics. Guidance issued by the World Health Organization in 2021 states that AI developers and users should:

> Ensure that all persons are treated fairly, which includes the requirement to ensure that no person or group is subject to discrimination, neglect, manipulation, domination or abuse (sometimes called 'justice' or 'fairness').[96]

However, the guidance makes no attempt to explain how this principle might be implemented. Similarly UNESCO's 2021 Recommendation merely asserts without further analysis that 'AI actors should promote social justice and safeguard fairness and non-discrimination of any kind in compliance with international law' and that 'AI actors should make all reasonable efforts to minimize and avoid reinforcing or perpetuating discriminatory or biased applications and outcomes throughout the life cycle of the AI system to ensure fairness of such systems.'[97]

A deeper and more extensive analysis of the ethics of AI, including fairness, is found in the 2019 roadmap produced jointly by the Nuffield Foundation and

[92] Mittelstadt (n 89).
[93] House of Lords Select Committee on Artificial Intelligence, *AI in the UK – ready, willing and able?* (2018 HL Paper 100).
[94] ibid, paras 349–65, 395–401.
[95] ibid, para 417.
[96] World Health Organization, *Ethics and governance of artificial intelligence for health: WHO guidance* (2021) 23, www.who.int/publications/i/item/9789240029200.
[97] UNESCO, *Recommendation on the Ethics of Artificial Intelligence* (2021) paras 28 and 29, unesdoc.unesco.org/ark:/48223/pf0000380455.

the Leverhulme Centre for the Future of Intelligence at the University of Cambridge.[98] This reviews the developing consensus on the ethical issues which AI generates, identifying the very different understandings of ethical concepts across a range of disciplines, primarily computer science, qualitative social science, philosophy, policy and law. Fairness is as equally contested as all other ethical issues. The authors do not attempt to set out a set of ethical principles, but instead analyse a range of different approaches which developers might take in the context of AI. Two important points become clear from this work:

- Each possible approach involves a number of societal, economic and technical trade-offs, indicating that there are no simple solutions which can be implemented across the board for AI; and, as a consequence
- Developers of AI need to make difficult choices about AI fairness, as do law and policy makers.[99]

Some analytical work attempts to explain how fundamental ethical principles such as fairness might be translated into practice by AI developers and users. The guidelines produced by the EU High Level Expert Group on Artificial Intelligence[100] explain the kinds of choices which need to be made for fairness and bias avoidance and set out some high-level principles to guide those choices.[101] The guidelines conclude with a set of questions which developers and users should address, including:

> Did you ensure an adequate working definition of 'fairness' that you apply in designing AI systems?
>
> - Is your definition commonly used? Did you consider other definitions before choosing this one?
> - Did you ensure a quantitative analysis or metrics to measure and test the applied definition of fairness?
> - Did you establish mechanisms to ensure fairness in your AI systems? Did you consider other potential mechanisms?[102]

This reinforces the finding from chapter three that an important route to achieving AI fairness lies in articulating and justifying the choices made.

[98] Jess Whittlestone, Rune Nyrup, Anna Alexandrova, Kanta Dihal and Stephen Cave, *Ethical and societal implications of algorithms, data, and artificial intelligence: a roadmap for research* (Nuffield Foundation and Leverhulme Centre for the Future of Intelligence, University of Cambridge, 2019), www.nuffieldfoundation.org/wp-content/uploads/2019/12/Ethical-and-Societal-Implications-of-Data-and-AI-report-Nuffield-Foundat.pdf.

[99] See also Toby Walsh, Neil Levy, Genevieve Bell, Anthony Elliott, James Maclaurin, Iven Mareels and Fiona Wood, *The effective and ethical development of artificial intelligence: An opportunity to improve our wellbeing* (2019 Report for the Australian Council of Learned Academies) 179, available at www.acola.org, making a similar point about choices.

[100] High Level Expert Group on Artificial Intelligence, *Ethics Guidelines for Trustworthy AI* (European Commission, 2019), digital-strategy.ec.europa.eu/en/library/ethics-guidelines-trustworthy-ai.

[101] ibid, 12 and 18.

[102] ibid, 30.

Leslie, in a guide published by the Alan Turing Institute in 2019,[103] sets out a fundamental ethical principle of fairness:

> [T]he principle of discriminatory non-harm as a minimum required threshold of fairness. This principle directs us to do no harm to others through the biased or discriminatory outcomes that may result from practices of AI innovation.[104]

He then translates this into four further principles, proposing that AI developers and users should ensure that AI:

1. Are trained and tested on properly representative, relevant, accurate, and generalisable datasets (Data Fairness)
2. Have model architectures that do not include target variables, features, processes, or analytical structures (correlations, interactions, and inferences) which are unreasonable, morally objectionable, or unjustifiable (Design Fairness)
3. Do not have discriminatory or inequitable impacts on the lives of the people they affect (Outcome Fairness)
4. Are deployed by users sufficiently trained to implement them responsibly and without bias (Implementation Fairness)

The guide contains a detailed discussion of the issues which face AI producers in seeking to comply with these principles, including in-depth discussion of different technical approaches. Throughout, Leslie stresses that producers are making choices about the meaning they attach to fairness and the mechanisms they choose to adopt to implement that meaning. No 'right' answer is possible. However, what is possible in each case is a *justifiable* answer, and thus he recommends:

> [A] Fairness Position Statement (FPS) in which the fairness criteria being employed in the AI system is made explicit and explained in plain and non-technical language. This FPS should then be made publicly available for review by all affected stakeholders.[105]

As the EU High Level Expert Group makes a similar recommendation, it might be concluded that although AI ethics is unable to translate ethical principles into rules for AI development and use, nonetheless ethical principles are the touchstone against which any justification for AI fairness choices should be judged. Ethical compliance ought thus not to be seen as a rule-following exercise, but rather the identification and selection among ethically acceptable choices and, most importantly, convincing society that those choices were justifiable.[106]

[103] David Leslie, *Understanding artificial intelligence ethics and safety: A guide for the responsible design and implementation of AI systems in the public sector* (Alan Turing Institute, 2019), www.turing.ac.uk/sites/default/files/2019-06/understandingartificialintelligenceethicsandsafety.pdf.

[104] ibid, 14.

[105] ibid, 20.

[106] At least one major AI developer and user seems to agree. GCHQ, a branch of the UK security services, writes in GCHQ, *Pioneering a New National Security The Ethics of Artificial Intelligence* (2021) 28, www.gchq.gov.uk/artificial-intelligence/index.html, that:

> At GCHQ, we will be able to demonstrate that we have implemented measures to minimise bias, and that we have scrutinised our AI models' outputs to mitigate the potential for discriminatory outcomes. We have extensive expertise in applying complex analytics and mathematical models to large datasets, and will utilise both this and wider best practice to develop this approach.

This approach to AI ethics and fairness departs from the top-down approach, and is much closer to bottom-up development of AI ethics as proposed by Mittelstadt[107] and John-Mathews et al.[108]

4.1.2. Fairness and Equality

If fair AI involves making choices, we clearly need to understand the reasoning behind the different fairness approaches which can be adopted. Philosophy provides extensive guidance here, though no clear consensus.

Philosophers seem to agree that fairness involves some element of resource reallocation to achieve equality. Even those who deny any need for action to achieve equality at the societal level[109] agree that fairness in individual decisions involves treating people equally.

The groundwork for the philosophical analysis of fairness was set by John Rawls in his *Theory of Justice*.[110] In that book he proposes high-level principles on which a just society should be based, conceiving justice in terms of fairness. It is important to recognise that his analysis explains how a society should be structured and organised such that there is a fair distribution of opportunity and benefits, rather than defining the characteristics of a fair decision about allocation in individual cases.[111] However, elements of his theory have been applied to individual decision-making, particularly by researchers into AI fairness.

The general outline of Rawls' theory is summarised as follows:

> All social primary goods – liberty, opportunity, income and wealth, and the bases of self-respect – are to be distributed equally unless an unequal distribution of any or all of these goods is to the advantage of the least favored.[112]

This is to be achieved by adopting two principles:[113]

> First Principle
>
> Each person is to have an equal right to the most extensive total system of equal basic liberties compatible with a similar system of liberty for all.

However, there does not yet appear to be any published statement by GCHQ explaining, at least at a non-detailed level, what AI fairness techniques have been adopted and why these are thought to achieve fairness.

[107] Mittelstadt (n 89).

[108] John-Mathews et al (n 91).

[109] See eg Friedrich Hayek, *The Mirage of Social Justice* (Chicago, University of Chicago Press, 1976).

[110] John Rawls, *A Theory of Justice* (Cambridge, MA, Harvard University Press, 1971).

[111] Throughout, Rawls makes it clear that his is a theoretical societal structure which overall aims at producing fair allocations, rather than guaranteeing them in individual cases – see eg his adaptation of Pareto optimality, ibid 66–67. See also ibid 304: 'When the basic structure takes this form the distribution that results will be just (or at least not unjust) whatever it is.'

[112] ibid, 303.

[113] The argument is based on an assumption that the members of a society, ignorant of their own position in that society and their relative advantages and disadvantages, would choose these organising principles in preference to any alternatives – ibid, 14–15, 31.

> Second Principle
>
> Social and economic inequalities are to be arranged so that they are both:
>
> (a) to the greatest benefit of the least advantaged … and
> (b) attached to offices and positions open to all under fair equality of opportunity.[114]

The second principle guides individual allocative decisions, based primarily on equality of opportunity which means that all should have an equal chance of gaining the resource in question. Inequalities of opportunity should only exist if they enhance the opportunities of those who have the lowest opportunity, and the societal structure should allow for sufficient accumulation of capital,[115] in the form of individual wealth, physical and social infrastructure, and so on, to enhance the prospect of increased opportunity in the future for those with the lowest opportunity. This does not lead to equal shares for all, or even the highest possible minimum share for those who are least advantaged. Rather, each member of society receives the share that meets their legitimate expectations,[116] expectations which have been generated by establishing a society based on these principles.

Carey and Wu[117] explain how this theory of justice as fairness has been translated into machine learning approaches for the development of AI. The focus here is on ensuring equality of the opportunity to achieve a favourable decision from the AI, and is broadly undertaken in two ways.

The first is in terms of 'substantive' equality of opportunity. This derives from Rawls' conception of fairness and justice, and assumes that talent is equally distributed among society. As a consequence, the AI's decision-making requires weighting to counter educational and environmental disadvantage which prevents the decision subject from demonstrating that talent. However:

> [T]he assumption that these talents are equally distributed often does not hold in practice. By the time a machine learning system is being used to make a decision, it is normally too late to provide individuals with the opportunity to develop qualifications. In this lens, fair machine learning has re-written Rawls' [equality of opportunity] to say that a competition must only measure a candidate on the basis of their talent, while ignoring qualifications that reflect the candidates unequal developmental opportunities prior to the point of the competition.[118]

The second approach is one of 'luck-egalitarian' equality of opportunity. This assumes 'that a person's outcome should be affected only by their choices, not their circumstances', but suffers because 'the difficulty of separating choice from circumstance is non-trivial, and in practice, quite difficult to implement.'[119] The proposed solutions (see chapter seven) attempt to control for disadvantaging factors such as race and poverty on an individual basis, often by weighting factors

[114] ibid, 302.
[115] This is what Rawls describes as the 'just savings principle', ibid, 284–97.
[116] ibid, 313–14.
[117] Alycia N Carey and Xintao Wu, 'The Fairness Field Guide: Perspectives From Social And Formal Sciences' (2022) *AI and Ethics* 1.
[118] ibid, 5.
[119] ibid, 5.

like qualifications in an attempt to compensate for disadvantage. However, making these individual adjustments means that it is no longer possible to make comparisons between sub-groups to demonstrate that the AI is achieving the overall, collective fairness which Rawlsian theory demands.

4.1.3. Luck Egalitarianism

The luck egalitarian modifications to the Rawlsian conception of fairness seem to derive in large part from Dworkin. Dworkin begins with the same starting point as Rawls, that fairness derives from equal distribution, but then analyses what this means in terms of individuals as well as for society as a whole.

Dworkin's theory is set out in two articles from 1981. The first identifies that there are two, competing, conceptions of equality: distribution to achieve equality of welfare (which includes equality of opportunity); and distribution merely to achieve equality of resources.[120] This article argues that aiming at equality of welfare is a politically undesirable goal, primarily because even if society is able to agree on a choice between the two possible bases of such equality, neither can be applied without reference to some external principle for making choices about their meaning and application.

One possible basis of equality of welfare is to redistribute so that everyone has equality of success in achieving their individual preferences. However, each individual makes their own choices, and values their own preferences differently, so that there can be no objective measure of success.[121] Dworkin suggests that the only reasonable test for equality is that everyone has a similar absence of reasonable regret about the share they have received:

> No one can reasonably regret that he has not had the life that someone with supernatural physical or mental powers, or the life span of Methuselah, would have had. So no one has a less successful life, all things considered, just because he thinks that such a life would be infinitely more valuable, in the philosophical way, than the

[120] Ronald Dworkin, 'What is Equality? Part 1: Equality of Welfare' (1981) 10 *Philosophy & Public Affairs* 185, 186:

> I shall consider two general theories of distributional equality. The first (which I shall call equality of welfare) holds that a distributional scheme treats people as equals when it distributes or transfers resources among them until no further transfer would leave them more equal in welfare. The second (equality of resources) holds that it treats them as equals when it distributes or transfers so that no further transfer would leave their shares of the total resources more equal.

[121] ibid, 208:

> A person of very limited talents might choose a very limited life in which his prospects of success are high because it is so important to be successful at something. Another person will choose almost impossible goals because for him the meaning is the challenge. Equality of relative success proposes to distribute resources – presumably much fewer to the first of these two and much more to the second – so that each has an equal chance of success in meeting these very different kinds of goals.

> life he has. But people can reasonably regret that they have not had the normal powers or the normal span of life that most people have. No one can reasonably regret that he has not had the life that someone with an unfairly large share of the world's resources would have led, so no one person's life is less successful than another's because the first thinks his life would be much more valuable in those circumstances while the other does not. But people can reasonably regret not having whatever share of material resources they are entitled to have.[122]

Regret is, of course, equally as subjective a measure as welfare, and cannot practicably be used as the basis for making distributions.

A second basis for equality of welfare might be equality in terms of each person's balance of enjoyment and satisfaction, but again this is completely subjective and cannot be used as a measure of allocation success.[123]

This leads Dworkin to the conclusion that any objective test for determining equality of welfare requires some theory of fair distribution to justify it, and that this theory cannot itself be based on equality of welfare.[124] As a consequence, the only possible basis for fair distribution is that of equality of resources.

Dworkin's second article[125] analyses equality of resources by positing a hypothetical market via which resources can be traded, and a hypothetical insurance system which allows individuals to insure against risks, including the risk of having a low income. Each individual ought to begin with the same value of resources as every other,[126] thus demanding that those who begin with disadvantages such as physical or mental disabilities should receive extra resources to compensate for the lower value of their physical or mental abilities. The test for success is, as for equality of welfare, the equally difficult one of regret: 'No division of resources is an equal division if, once the division is complete, any [individual] would prefer someone else's bundle of resources to his own bundle.'[127]

Thereafter, though, inequality begins to creep in. Some individuals work harder or more successfully than others; some experience good luck and some bad. Some, but not all,[128] of the consequent inequalities lead to immediate unfairness which should be compensated for as that unfairness is recognised.

[122] ibid, 216.

[123] ibid, 221.

[124] ibid, 221–22.

[125] Ronald Dworkin, 'What is Equality? Part 2: Equality of Resources' (1981) 10 *Philosophy & Public Affairs* 283. A few years later Dworkin examined other, non-distributional aspects of equality which are not relevant to this discussion – see Ronald Dworkin, 'What is equality Part 3: The place of liberty' (1987) 73 *Iowa Law Review* 1; Ronald Dworkin, 'What is equality Part 4: Political equality' (1987) 22 *University of San Francisco Law Review* 1.

[126] Recognising the difficulties of achieving this, because of the different subjective values each individual places on different assets. His thought experiment of a group of shipwrecked travellers on a deserted island, auctioning off their resources using an equal allocation of otherwise worthless shells, is an ingenious solution for this simple society, but unsurprisingly Dworkin does not attempt to describe a workable system for more complex societies.

[127] Dworkin (n 125) 285.

[128] Dworkin distinguishes option luck, where an individual's choices either do or do not lead to the desired consequences, from brute luck such as unpreventable illness or accident. The envy test which

Dworkin also addresses the long-term consequences of growing inequality. He makes it clear that he does not subscribe to:

> [W]hat we might call the starting-gate theory of fairness: that if people start in the same circumstances, and do not cheat or steal from one another, then it is fair that people keep what they gain through their own skill. But the starting-gate theory of fairness is very far from equality of resources. Indeed it is hardly a coherent political theory at all. The starting-gate theory holds that justice requires equal initial resources. But it also holds that justice requires laissez-faire thereafter, in accordance, presumably, with some version of the Lockean theory that people acquire property by mixing their labor with goods or something of that sort. But these two principles cannot live comfortably together. Equality can have no greater force in justifying initial equal holdings … than later in justifying redistributions when wealth becomes unequal because people's productive talents are different.[129]

He argues that this inequality ought not to be compensated for by individual distributions to equalise resources, because as a matter of principle 'if people of equal talent choose different lives it is unfair to redistribute halfway through those lives.'[130] This concept is often termed luck egalitarianism, in which individuals are compensated for inequalities arising from unavoidable misfortune but not for inequalities which are the consequences of the voluntary choices they make.[131]

To deal with the wider social inequality which arises from luck egalitarianism, Dworkin argues that collective redistribution should take place via the tax system of that society. This:

> [L]eaves intact the possibility of choosing a life in which sacrifices are constantly made and discipline steadily imposed for the sake of financial success and the further resources it brings, though of course it neither endorses nor condemns that choice. But it also recognizes the role of genetic luck in such a life. The accommodation it makes is a compromise; but it is a compromise of two requirements of equality, in the face of both practical and conceptual uncertainty how to satisfy these requirements, not a compromise of equality for the sake of some independent value such as efficiency.[132]

However, taxation only reduces inequality, rather than eliminating it, thus leaving scope for envy. This envy on its own does not justify further redistribution, which would require justification on some other basis than mere inequality.[133]

determines fairness says that individuals should be compensated through redistribution for brute luck losses, but not for the option luck outcomes resulting from their choices – ibid, 293–99.

[129] ibid, 309.

[130] ibid, 310.

[131] See eg Thomas Nagel, *Equality and Partiality* (New York, Oxford University Press, 1991); Eric Rakowski, *Equal Justice* (New York, Oxford University Press, 1991).

[132] Dworkin (n 125) 312.

[133] ibid, 334:

> Equality of resources is a complex ideal. It is probably (as the various arguments we have canvassed in this essay suggest) an indeterminate ideal that accepts, within a certain range, a variety of

One difficulty with applying Dworkin's ideas to AI development lies in identifying the starting point at which we should ensure initial equality. Carey and Wu's research[134] tells us that AI developers have assumed they should apply this test at the time the AI is making its decision, or in other words that one aspect of fair AI is that it should redistribute the resources it is allocating in a way which reduces current levels of inequality.

A second difficulty is whether AI developers should follow Dworkin and redistribute only to equalise resources (including negative resources such as lack of physical or mental ability, lack of educational opportunities, etc), or should equalise for disadvantages which arise from the choices made by individuals in an attempt to achieve equality of welfare, or should adopt some other approach. Arneson reviews the possible approaches to egalitarianism and identifies three conceptions of equality which are relevant to our discussion:[135]

- Formal equality of opportunity, which is particularly relevant to decisions allocating matters such as employment, licences to undertake an activity and so on. This is achieved if the decision is based on the objective qualities of the decision subject, avoiding irrational biases or prejudices. This conception would therefore exclude the possibility of reverse discrimination in favour of disadvantaged groups or individuals, to compensate for their inability to acquire the necessary objective qualities.
- Fair equality of opportunity, based on Rawls. In this conception, decisions should be based on the innate qualities of the decision-subject, ignoring those qualities which have been acquired because of inequality of resources such as parental incomes to fund training and education. The difficulties in this conception are obvious, as most qualities relevant to, for example, an employment decision are not observable directly, but rather are perceived through secondary indicators such as qualifications and experience. And, as Arneson points out, some apparently innate qualities can arise as a result of inequality.[136] Implementing this conception of equality in a decision-system involves political decisions about how society *ought* to have been organised more equally.[137]

different distributions. But this much seems clear: any defensible conception of that ideal must attend to its different dimensions, and not reject out of hand the requirement that it be sensitive to the cost of one person's life to other people. The present suggestion, that genuine theories of equality must be concerned only with the quantity of disposable goods or liquid assets people command at a particular time, is a piece of pre-analytic dogma that does not, in fact, protect the boundaries of the concept of equality from confusion with other concepts, but rather thwarts the attempt to picture equality as an independent and powerful political ideal.

[134] Carey and Wu (n 117).

[135] Richard Arneson, 'Four conceptions of equal opportunity' (2018) 128 *The Economic Journal* F152. A fourth, libertarian equality of opportunity, is irrelevant to decision-making.

[136] One example, ibid F157, is the quality of ambition and work ethic, which can be trained into an individual through their expensive education.

[137] Arguably, wider society is both better placed to make those political decisions, and to act to counter the consequences of inequality via education and training, than an individual decision-maker – see ibid F159–60.

- Luck egalitarian equality of opportunity, already explained when discussing Dworkin's theory of fairness. Arneson points out[138] that the individual choices whose consequences are to be ignored are themselves socially determined by upbringing and education, which in turn are heavily influenced by inequality. This makes it impossible to determine how far any choice is in fact voluntary,[139] and thus means that luck egalitarianism cannot be viewed as a principle for achieving equality.

Arneson's conclusion is thus that all three of these are normative mechanisms which society might choose to use in furtherance of some other egalitarian principle, but do not themselves describe any such principle.

Roemer[140] attempts to avoid all these problems by moving the starting point for assessing equality of opportunity as far back as possible. According to his conception of equality of opportunity, equality is achieved if those individuals who try equally hard can achieve equivalent outcomes. Thus when comparing two individuals it would be necessary to assess their starting points and potential for achievement, from which it can be determined whether they have put in equivalent effort. If so, we make decisions so that each achieves an equivalent outcome:

> [F]rom a moral viewpoint, we declare that two individuals in different types have tried equally hard if they lie at the same rank of the effort distributions of their types, and our policy aims to render such persons equal in final condition, independent of their circumstances.[141]

Metrics are developed for the purposes of such assessments, and resources are allocated in accordance with those metrics. This aims to avoid the problem of identifying which choices are truly voluntary and which are determined by the individual's circumstances, assuming that failure to achieve full potential according to the relevant metric can only be the result of deliberate choice to expend effort elsewhere, or to make less effort. In theory if any individual then falls short on outcomes, that is because they put in less effort than they are capable of. 'To the extent that differences in outcome are due to circumstance, it approaches Rawlsianism, but to the extent they are due to effort, it approaches utilitarianism.'[142]

This conception of equality might be usable for decisions in fields like employment, both because effort is a relevant reason for preferring one candidate over

[138] ibid, F166.
[139] ibid, F172:

> On a fine-grained view, it might turn out that I am more to blame for ordinary bourgeois faults such as stupidly becoming drunk and misbehaving at my brother's wedding than some axe murderers are for their bloody deeds, pushed on them by overpowering desires and resisted by them to an almost heroic extent. To reiterate, even if these notions belong in the formulation of fundamental principles, if any feasible attempts to implement them in given circumstances would be counterproductive, the luck egalitarian principles themselves counsel against any attempt at implementation.

[140] John Roemer, 'Equality of opportunity: A progress report' (2002) 19 *Social Choice and Welfare* 455.
[141] ibid, 458.
[142] ibid, 460.

another and because metrics might be calculable. As an example, there is some evidence that metrics can be calculated to assess the potential for degree education success of students from disadvantaged backgrounds with low school grades, which could be used to decide if their effort in achieving those grades was equivalent to higher-scoring students from other backgrounds and thus deserved university admission.[143] However, the effort expended by a patient usually plays little or no part in decisions about the allocation of medical treatment to them,[144] and so Roemer's conception would not be usable here.

We thus see that although there is clear agreement that fairness is based on equality, there is no philosophical consensus on how best to *achieve* equality. Luck egalitarianism has proved attractive to AI researchers but it is a contested concept, trenchantly criticised by Anderson as departing from the

> proper negative aim of egalitarian justice [which] is not to eliminate the impact of brute luck from human affairs, but to end oppression, which by definition is socially imposed … [and] not to ensure that everyone gets what they morally deserve, but to create a community in which people stand in relations of equality to others.[145]

This broader aim is outside the scope of individual AI decisions, of course, and therefore ignored in AI development, as is the more general redistribution which Dworkin advocates. All these matters must be left to the institutions of wider society.

4.1.4. Welfare Egalitarianism

Other philosophical theorists reject luck egalitarianism in favour of welfare egalitarianism, attempting to counter the objections of Rawls and Dworkin. Welfare can be measured either in terms of the degree of pleasure which individuals

[143] Vikki Boliver, Stephen Gorard and Nadia Siddiqui, 'Will the use of contextual indicators make UK higher education admissions fairer?' (2015) 5 *Education Sciences* 306.

[144] Even here there are exceptions, where patients are required to change their lifestyle (losing weight, giving up smoking, and so on) before they will receive treatment. However, the granting of treatment is not a reward for effort per se, but rather because without that expenditure of effort the patient is less likely to benefit from the treatment.

[145] Elizabeth S Anderson, 'What is the Point of Equality?' (1991) 109 *Ethics* 287, 288. She argues (p 69) that it fails in three ways:

> First, it excludes some citizens from enjoying the social conditions of freedom on the spurious ground that it's their fault for losing them. It escapes this problem only at the cost of paternalism. Second, equality of fortune makes the basis for citizens' claims on one another the fact that some are inferior to others in the worth of their lives, talents, and personal qualities. Thus, its principles express contemptuous pity for those the state stamps as sadly inferior and uphold envy as a basis for distributing goods from the lucky to the unfortunate. Such principles stigmatize the unfortunate and disrespect the fortunate by failing to show how envy can obligate them. Third, equality of fortune, in attempting to ensure that people take responsibility for their choices, makes demeaning and intrusive judgments of people's capacities to exercise responsibility and effectively dictates to them the appropriate uses of their freedom.

experience, or by how far their preferences are satisfied;[146] but it is arguable that these two effectively cover the same ground except at their extremes.[147] Welfare egalitarians respond to objections by seeking not equality of welfare per se, but rather equality of the *opportunity* for welfare, or equal access to advantage.

Arneson espouses equality of the opportunity for welfare. He takes welfare to mean preference satisfaction.[148] Equality of opportunity is achieved if each individual is in the same position as others to satisfy their preferences to an equivalent degree.[149] Of course, choices or negligent behaviour which reduce an individual's potential to satisfy their preferences thereby reduce their equality of opportunity for welfare, but the voluntariness of those choices or behaviours justifies society in not compensating them for the reduction in opportunity.[150]

He accepts, though, that in practice this approach to equality is impracticable for individual decisions. This is because the information necessary to decide which choices are voluntary is either unavailable or it is socially undesirable that it should be collected and used. This drives him back so something very close to Dworkin's conception: 'We may insist that governments have regard to primary good share equality or resource equality as rough proxies for the welfarist equality that we are unable to calculate.'[151]

Cohen argues that what society should be seeking to maximise is individual welfare, measured in terms of satisfying their pleasure or preference. Pleasure and preference are purely subjective – if one person's palate can only be satisfied by fine wine, whilst another's is equally satisfied with cheaper beer, the former is entitled to a greater share of resources so that each can be equally contented. This is subject to one caveat – that if a deliberate choice on the part of the individual reduces their access to advantage, that becomes the individual's own responsibility and thus society has no obligation to compensate for it.[152]

> I distinguish among expensive tastes according to whether or not their bearer can reasonably be held responsible for them. There are those which he could not have helped forming and/or could not now unform, and then there are those for which, by contrast, he can be held responsible, because he could have forestalled them and/or because he could now unlearn them. Notice that I do not say that a person who deliberately develops an expensive taste deserves criticism. I say no such severe

[146] GA Cohen, 'On the Currency of Egalitarian Justice' (1898) 99 *Ethics* 906, 911.

[147] For example, I might have an intellectual preference for complex and cerebral music, but actually achieve more pleasure from hearing a popular song.

[148] Arneson, 'Equality and Equality of Opportunity for Welfare' (1989) 55 *Philosophical Studies* 79, 82.

[149] ibid, 85–86.

[150] ibid, 86:

> [I]n an extended sense people share equal opportunity for welfare just in case there is some time at which their opportunities are equal and if any inequalities in their opportunities at later times are due to their voluntary choice or differentially negligent behavior for which they are rightly deemed personally responsible.

[151] ibid, 87.

[152] Cohen (n 146) 924.

> thing because there are all kinds of reasons why a person might want to develop an expensive taste, and it is each person's business whether he does so or not. But it is also nobody else's business to pick up the tab for him if he does. Egalitarians have good reason not to minister to deliberately cultivated expensive tastes, and equality of welfare must, therefore, be rejected. But we should not embrace equality of resources instead, since that doctrine wrongly refuses compensation for involuntary expensive tastes, and it does not refuse compensation for voluntary ones for the right reason.

Cohen's conception of fair allocation is clearly linked to luck egalitarianism:

> In my view, a large part of the fundamental egalitarian aim is to extinguish the influence of brute luck on distribution. Brute luck is an enemy of just equality, and, since effects of genuine choice contrast with brute luck, genuine choice excuses otherwise unacceptable inequalities.[153]

Although in theory Cohen's concept could be used for individual decision-making, in practice it seems impossible to implement. Pleasure and preference are purely subjective and thus can only be self-certified by the individual in question. Identifying deliberate choices, as opposed to apparent choices which are in fact determined by external factors such as upbringing, is even harder. Indeed, Cohen accepts that this lands us 'up to our necks in the free will problem',[154] but says this is unfortunate but inevitable if we base our egalitarianism on choices.

4.1.5. Philosophical Guidance for AI Developers

This review of philosophical conceptions of fairness is far from exhaustive, but even so it identifies a dauntingly wide menu of ideas which are not all compatible with each other. Can we find any kind of consensus?

There are two themes which most philosophical thinkers about fairness seem to agree with:

- Individual decisions should be based on equal treatment of decision subjects; and
- Equal treatment requires some form of weighting for disadvantage, without which there is a risk that the advantaged will be more likely to receive a favourable decision than the disadvantaged.

The disagreements are largely about how disadvantage should be accounted for, linked to ideas of responsibility. In particular, luck egalitarianism takes the position that only innate disadvantage should receive more favourable treatment, while disadvantage resulting from deliberate choices is the responsibility of that individual and deserves no special consideration.

[153] ibid, 931.
[154] ibid, 934.

A second point of disagreement is about whether disadvantage should be compensated for at a single point in time, leaving all with equal opportunities to achieve if they make good choices and exert equivalent effort, or whether all allocative decisions require appropriate compensation for the current state of disadvantage (with substantial disagreement about what is appropriate). It is worth noting that the first group aims at an idealistic view of social justice, an initial state of equality, which it is doubtful that any human society has in fact achieved. This leaves AI developers with two choices. First, they might decide that it is for wider society to achieve equality of advantage through positive measures such as benefits and education, so that their AI should simply seek to achieve equality on objectively relevant metrics such as qualifications, health data, and so on. Alternatively they might decide that disadvantage which reduces an individual's score on those relevant metrics should be excluded from influencing the decision, most likely by applying suitable weightings to the metrics affected – for example, in the scoring of a job application evaluation AI, a university degree obtained by an individual from a deprived background might be weighted more highly than the same degree awarded to a privileged individual.

Developers who adopt the second approach face the difficulty of identifying appropriate weightings, but the work of Le Franc et al[155] suggests one way in which this might be done. They attempt to construct a model of equality of opportunity which incorporates as many elements as possible of the disparate philosophical theories, and identify three main elements to be incorporated: 'What equality of opportunity requires is that, given effort, no one faces more favorable outcome prospects, as a result of luck, for reasons related to differential circumstances.'[156]

The circumstances which they account for are 'the deterministic or random factors that are not considered a legitimate source of inequality' by society.[157] Effort is assumed to be a legitimate source of inequality, as is luck which they define as 'the random factors that are seen as a legitimate source of inequality as long as they affect individual outcomes in a neutral way, given circumstances and effort.'[158]

How each of these is to be defined is a matter for political and social debate in each society, which therefore leaves AI developers to seek indications of political and social consensus from external sources. These might come from law and regulation, such as legal definitions of unlawful discrimination, or from ethical or regulatory recommendations (see section 5.3.1 and section 4.1.1 above).

[155] Arnaud Lefranc, Nicolas Pistolesi and Alain Trannoy, 'Equality of opportunity and luck: Definitions and testable conditions, with an application to income in France' (2009) 93 *Journal of Public Economics* 1189.
[156] ibid, 1190.
[157] ibid, 1192.
[158] ibid, 1192.

Once definitions have been decided, weightings for metrics can be identified (in some cases at least) by analysing data about inequality in society as a whole. This can produce an average measure of disadvantage for particular types of individual which can then be used to calculate weightings.

> All conceptions of equality of opportunity discussed so far amount to equalize the prospects for outcome that individuals face, given their effort level. These prospects obviously refer to the ex ante distribution of outcomes, before the individuals know their degree of luck. Of course this ex ante distribution is never observable empirically. At best, what we can recover is the ex post distribution of outcome among individuals of a given variety. The empirical characterization of equality of opportunity undertaken in the previous sections rests on the assumption that the ex ante prospects offered to individuals of a given variety can be identified with the ex post distribution of outcomes among individuals of that type. In other words, what is required is a law of large numbers implying that the frequency distribution in the cross section be identical to the probability distribution for the individual.[159]

The authors illustrate this process via an economic model based on these concepts, and use it to analyse employment inequality in France via income data for different categories of employees, comparing the years 1979 and 2000. They find that inequality of opportunity persists, but that the differentials have decreased over time, and from this numerical assessment of inequality it would clearly be possible to calculate weightings which, if applied in employment decisions, should lead to equality on average.

In my view, we end up in the position that the philosophy of fairness offers AI producers a range of fairness tests from which they must select. That selection is likely to be driven by the external demands of society, but even more by the decision-making aims of the AI. Decision-making aims can be set by law (such as by prohibiting some kinds of discrimination), by the need for societal acceptance (our medical treatment allocation AI might be an example), by the needs of the AI user (to employ the applicant with the highest potential to do the job well), and so on. Philosophy does not tell developers the 'right' answer about fairness, though it does explain what different fairness techniques are maximising for. Whether its conceptions are usable depends very much on data availability. We have seen that satisfaction, preference and deliberate choice are all fairness criteria which make sense as part of a philosophical argument, but for which reliable data on individuals is at present unlikely to be available.[160]

[159] ibid, 1196.

[160] We should note, though, that research in sentiment analysis is advancing rapidly, and that some insight into these matters can be achieved from data such as social media postings – see Lei Zhang, Shuai Wang and Bing Liu, 'Deep learning for sentiment analysis: A survey' (2018) 8 *Wiley Interdisciplinary Reviews: Data Mining and Knowledge Discovery* e1253.

4.1.6. Philosophy and the Social Conception of Fairness

If we look back at the discussion of social fairness in chapter three it is apparent that the common factor with philosophical approaches to fairness is that each seeks equality. The primary difference is that the social conception looks for equality of outcome whilst philosophy concentrates on equality of opportunity. How significant is this difference?

In most cases requiring a decision it seems likely that either conception would lead to roughly the same answer via different routes. An equality of opportunity approach to decision-making treats differential outcomes as fair if each individual had the same opportunity to succeed, making appropriate adjustments to correct for innate disadvantage. The social conception considers unequal outcomes to be unfair unless differences are justified on the basis of merit and disadvantage, balanced against each other. A similar balancing of merit and disadvantage is being undertaken, but from different time perspectives. Social fairness looks at the situation right now, assessing current merit and disadvantage. Philosophical fairness looks back to a time when the individuals under consideration were equal, or should have been equal, and considers changes in merit and disadvantage since then with adjustment to account for the effects of innate disadvantage.

An example might help to clarify this. Suppose that our medical treatment allocation AI is deciding between two patients. Patient A has a 70 per cent chance of recovering fully if treated, while Patient B has only a 50 per cent chance. However, Patient B suffered from childhood illness due to family poverty, which in part caused them to be less receptive now to the medical treatment. In adulthood Patient B had a successful career and is now as well off as Patient A, but became a non-addicted cocaine user (with no link to their childhood poverty) which explains their remaining level of lower receptivity.

Most philosophical approaches to fairness would suggest some adjustment to Patient B's fitness for treatment score to compensate for the childhood illness, but would treat the cocaine usage as a voluntary decision and not compensate for that. As a result, Patient A would receive the treatment. A social fairness approach would probably reach the same conclusion, compensating for childhood disadvantage, but treating the cocaine usage as a demerit which prevents the two cases from being equal.

If we change the facts slightly, so that both Patient A and Patient B have an equal 70 per cent chance to benefit, but Patient B's chances would have been higher without the childhood illness but have been reduced to 70 per cent by the cocaine use, we might reach a different answer. From a philosophical equality of opportunity perspective the cases are now equal. From a social fairness perspective, all depends on how that society's values perceive cocaine use. It might be seen as morally neutral, in the same way as excess consumption of alcohol is provided that it only harms the drinker. If so, A and B would rank equally in fairness terms. Or it might be seen as morally bad, because cocaine use fosters drug trafficking which involves exploitation of vulnerable children,[161] which would lead to a finding that it was fair that Patient A should be preferred to Patient B.

Additionally, the social conception of fairness takes account not just of disadvantage but also of need. In chapter three these were treated as broadly equivalent, but there may be cases where they should be distinguished. Fleurbay offers the example of a motorcycle accident victim whose injuries are more severe because he voluntarily decided not to wear a helmet with full knowledge of the risks this entailed.[162] There is no social medical system, and thus if the motorcyclist cannot afford the treatment for his injuries, Fleurbay reasons that an equality of opportunity approach will allow him to die because no additional allocation of resources is due to compensate for the adverse consequences of one's own decisions. Fleurbay argues[163] that this offends our moral sense such that we would disagree with the outcome which philosophy recommends. We noted in chapter three that the socially fair balance between merit and need/disadvantage changes with context, so from that perspective we might think that the context is now one in which overwhelming and urgent need outweighs any lack of merit.

We also noted in section 3.1.1 that whether differentials in individual outcome based on power and status are unfair is culturally determined. These factors play no part in the philosophical conception of fairness. However, I argued there that AI developers might reasonably decide to control for such inequality, perhaps even if such equalities are embedded as a societal consensus via law and regulation,[164] so this may make no practical difference for fair AI development.

This discussion leads to two possible interim conclusions. The first is that the choice between philosophical approaches to fairness which AI developers have to make should be influenced by the culture of the society in which the AI is intended to operate, and should also be influenced by ethical principles. Developers thus need to test the AI's outputs across a wide range of potential contexts, particularly the extremes of context, to identify if its decisions meet the selected

[161] See eg Grace Robinson, Robert McLean and James Densley, 'Working county lines: Child criminal exploitation and illicit drug dealing in Glasgow and Merseyside' (2019) 63 *International Journal of Offender Therapy and Comparative Criminology* 694.

[162] Marc Fleurbaey, 'Equal opportunity or equal social outcome?' (1995) 11 *Economics & Philosophy* 25, 40.

[163] ibid, 41–42.

[164] Law and regulation are notoriously slow to change, and so can embed conceptions of inequality which no longer match the social consensus. Additionally, even if law and regulation attempts to define some socially accepted aspects of fairness, implementation leaves many ethical choices to AI developers. Lee et al (n 31) 535, write:

> [O]ur position is that computer scientists and model developers cannot completely delegate this consideration to a third party, whether it is the regulator, business leader, or the risk function. Model developers must be engaged in the discussion on what layers of inequality should and should not be influencing the model's prediction in order to inform their decisions on model design, feature selection, and performance metric selection.

fairness tests. Thus the testing dataset for our hypothetical medical treatment allocation AI should include voluntary drug users, victims of avoidable injury, and such other outliers as the developers can envisage.

Second, in the same way that I recommend in section 8.4 that AI producers should explain how their fairness choices meet the social conception, they should also explain the aims behind the philosophically-derived fairness techniques they have implemented in their AI and justify those choices in terms of their likelihood to achieve fairness and meet ethical principles.[165] If a plausible justification cannot be constructed, it is likely that adopting those techniques will not result in a fair AI. Wong notes[166] that generating such explanations highlights the political disagreements about conceptions of fairness, and these are disagreements which can only be resolved through other mechanisms designed for the purpose such as law and regulation. Additionally, Peters argues that for humans the process of formulating an explanation for a decision has an element of 'mindshaping', which generates a normative impulse to follow that explanation when making future decisions.[167] Justifying fairness choices therefore enhances the chance that future decisions will be fair, at least in terms of the conception of fairness adopted for the explanation.

4.2. Economics of Fairness

4.2.1. Welfare Economics

Welfare economics aims to identify how overall social welfare can be improved without reducing overall utility. It thus considers social welfare primarily at the macro or societal level.[168] 'Welfare' in this context is used, broadly, in the

[165] The concept of Key Ethics Indicators, Lee et al (n 31) 539 ff, might be of assistance here, setting out a methodology for developing identifiable success metrics which could also be used to generate an ethical justification for developer choices.

[166] Wong (n 88), see particularly 230–31, 234–39.

[167] Uwe Peters, 'Explainable AI lacks regulative reasons: why AI and human decision-making are not equally opaque' (2022) *AI and Ethics* 1. Peters points out at 6 that an ex post explanation of an AI decision using XAI tools will have no influence on subsequent decisions:

> Crucially, the processing of the AI explanation system remains purely epistemic in nature: unlike in the case of human reason reports, in the case of ADM that is made intelligible by AI explanation models, there is no feedback between the rationalizing system and the decision-making system such that the former can alter the processing of the latter. AI explanation systems do not have future-directed, regulative effects on the black-box systems whose outputs they are to rationalize. They (currently) cannot manipulate the blackbox system's processing so that it aligns with the content of the rationalization.

[168] However, specific techniques for assessing social welfare might aggregate a number of analyses at the level of individual welfare – see eg Hoda Heidari, Claudio Ferrari, Krishna Gummadi and Andreas Krause, 'Fairness behind a veil of ignorance: A welfare analysis for automated decision making' (2018) 31 *Advances in Neural Information Processing Systems*.

Benthamite sense of overall pleasure or happiness,[169] but economists have necessarily to measure this via suitable proxies such as income or wealth. Where the welfare issue is the reduction of disadvantage, proxies are particularly difficult to define.[170]

Reallocation of resources is a primary mechanism for improving social welfare, and so welfare economics has concentrated on assessing the effects of any proposed or actual redistribution. A simplistic egalitarian redistribution in terms of assets, whether it equalises wealth or increases every individual's wealth by some amount, is usually rejected as inappropriate. This is because the welfare derived from assets varies between individuals – a sum of £100 might greatly increase the welfare of a poor person but make very little difference to the welfare of a rich person, and the same might be true of non-money assets like a loaf of bread except, to add further complexity, that the bread would reduce the welfare of someone with a wheat allergy.

This leads to a fundamental axiom of welfare economics, that any reallocation should be a Pareto-efficient solution. Pareto efficiency is achieved if, following the reallocation, resources cannot be further reallocated to make one individual better off without making at least one other individual worse off. This guarantees an increase in social welfare.

For any social welfare issue for which reallocation of resources is being considered, there are likely to be a number of different, Pareto-efficient ways in which the reallocation can be undertaken. In order to assess which of these produces the greatest social welfare, economists need to define a social welfare function. At one end of the spectrum we might find an egalitarian function for equal sharing, which is always a Pareto-efficient solution. At the other we have maxmin (also minimax) or Rawlsian utility functions, which define welfare as maximised when the benefit derived from a distribution which gives a greater share to those who initially have the least is also the greatest which can be achieve via a Pareto-efficient distribution. Between these extremes lie many other potential social welfare functions.

As a trivial but easy-to-understand illustration, imagine the sharing of a cake.[171] Any proposed method of sharing produces a Pareto-efficient solution because: (a) all now have more cake than before; and (b) further sharing of the cake would have to take cake from some to give it to others. An egalitarian, cake-sharing function would divide the cake into equal slices. A maxmin function might divide it unequally and give the largest slices to the hungriest.

[169] See Jeremy Bentham, *A Fragment on Government* (1776), Preface: 'it is the greatest happiness of the greatest number that is the measure of right and wrong'.

[170] See eg the detailed discussion of appropriate proxies for identifying discrimination in employment in Jonathan Guryan and Kerwin Kofi Charles, 'Taste-based or statistical discrimination: the economics of discrimination returns to its roots' (2013) 123 *The Economic Journal* F417.

[171] Though to game theorists, cake-cutting is far from trivial. For a complex analysis, involving truthfulness as well as fairness, see Yiling Chen, John K Lai, David C Parkes and Ariel D Procaccia, 'Truth, justice, and cake cutting' (2013) 77 *Games and Economic Behavior* 284.

Philosophical conceptions of fairness have proved attractive to economists because much disadvantage cannot be countered (and thus social welfare increased) merely by distributing assets. No amount of cake can overcome the disadvantages arising from race, sex, illness or a lack of education. Thus welfare economists have widely adopted the Rawlsian concept of fairness in terms of equality of opportunity, and sought to investigate how reallocating opportunity affects overall social welfare. They do this by developing economic models which can be used to measure how much improvement of equality of opportunity is likely to be achieved by any proposed technique for reallocation. The 'no-envy' test for fairness, which philosophers such as Dworkin have adopted, is seen as a core conception of welfare economics.[172]

A reliance on Rawls is particularly apparent in the work of economists when assessing the various methods proposed for achieving AI fairness. Here some variant of Rawlsian equality of opportunity usually forms the basis of the social welfare evaluations in their models. As we saw in sections 4.1.2–4.1.4, philosophers have understood equality of opportunity in multiple ways, and economists tend to adopt the understanding which best fits their preconceptions and their modelling. An analysis by Heidari et al shows that most existing notions of algorithmic fairness are special cases of equality of opportunity,[173] and the authors map economic models of equality of opportunity to different algorithmic fairness techniques to evaluate the differences between each technique's ways of measuring equality of opportunity.[174] Social welfare economic modelling is strongly influenced by Roemer's effort-based understanding of luck egalitarianism (see section 4.1.3).[175]

[172] The test suggests that a distribution is fair if no individual prefers another's share to their own – Duncan K Foley, 'Resource Allocation and the Public Sector' (1967) 7 *Yale Economic Essays* 45; Christian Arnsperger, 'Envy-Freeness and Distributive Justice' (1994) 8 *Journal of Economic Surveys* 155. For a review of modern conceptions of no-envy see Conrad Heilmann and Stefan Wintein, 'No Envy: Jan Tinbergen on Fairness' (2021) 14 *Erasmus Journal for Philosophy and Economics* 222.

[173] Hoda Heidari, Michele Loi, Krishna P Gummadi and Andreas Krause, 'A moral framework for understanding fair ML through economic models of equality of opportunity' (2019) *Proceedings of the Conference on Fairness, Accountability, and Transparency* 181, 185.

[174] ibid, 184.

[175] ibid, 181:

> Prior work in economics has sought to formally characterize conditions of equality of opportunity to allow for its precise measurement in practical domains. At a high level, in these models an individual's outcome/position is assumed to be affected by two main factors: his/her circumstance c and effort e. Circumstance c is meant to capture all factors that are deemed irrelevant, or for which the individual should not be held morally accountable; for instance c could specify the socio-economic status he/she is born into. Effort e captures all accountability factors – those that can morally justify inequality. (Prior work in economics refers to e as effort for the sake of concreteness, but e summarizes all factors for which the individual can be held morally accountable; the term "effort" should not be interpreted in its ordinary sense here.) For any circumstance c and any effort level e, a policy ϕ induces a distribution of utility among people of circumstance c and effort e. Formally, an EOP policy will ensure that an individual's final utility will be, to the extent possible, only a function of their effort and not their circumstances.

Analysis of AI fairness techniques via economic modelling identifies a number of criticisms of those techniques. Kasy and Abebe write:[176]

> [W]e argue that these leading notions of fairness, such as predictive parity or balance, suffer from the following three limitations.
>
> (1) They legitimize and perpetuate inequalities justified by 'merit' both within and between groups. The focus on 'merit' – a measure promoting the decision-maker's objective – reinforces, rather than questions, the legitimacy of the status quo.
> (2) They are narrowly-bracketed. Fairness only requires equal treatment within the context of the algorithm at hand, and does not consider the impact of the algorithm on inequality in the wider population. Unequal treatment that compensates pre-existing inequalities might reduce overall inequality.
> (3) They focus on categories (protected groups) and ignore within-group inequalities, e.g., as emphasized by intersectional critiques. Equal treatment across groups can be consistent with great inequality within groups.

Hu and Chen[177] criticise the broadness of most fairness measures, which generally focus on dividing the assessed population into two groups. 'Fairness constrained loss minimization offers little guidance on its own for choosing among the fairness desiderata, which appear incommensurable and result in different impacts on different individuals and groups.' They find in particular that applying a selected fairness measure more strictly, or less so, can make a substantial difference in welfare outcomes for groups not classified by the fairness measure. Additionally, a stricter application of the fairness measure often fails to achieve improved outcomes for the disadvantaged group.[178]

Other, recent economic work suggests that an AI fairness measure which simply measures the change in inequality between selected groups is an inadequate test for increased social welfare. This is because it fails to measure the overall effects which the measure would have in that society. Several authors have identified that although some implementations of fairness techniques might reduce inequality between the selected groups, viewed overall they in fact actually reduce total social welfare.[179] Mouzannar et al argue[180] that social welfare needs to be assessed over time to be accurate, not merely as a snapshot at the time of resource allocation, and they model how to identify which fairness techniques are

[176] Kasy and Abebe (n 44) 576.

[177] Lily Hu and Yiling Chen, 'Fair classification and social welfare' (2020) in *Proceedings of the 2020 Conference on Fairness, Accountability, and Transparency* 535.

[178] ibid, 535, 536.

[179] See Heidari et al (n 168); Hu and Chen (n 177); Lydia Liu, Sarah Dean, Esther Rolf, Max Simchowitz and Moritz Hardt, 'Delayed Impact of Fair Machine Learning' (2018) *Proceedings of the 35th International Conference on Machine Learning (Proceedings of Machine Learning Research)* Vol 80, 3156 proceedings.mlr.press/v80/liu18c.html; Sendhil Mullainathan, 'Algorithmic Fairness and the Social Welfare Function' (2018) *ACM EC'18: Proceedings of the 2018 ACM Conference on Economics and Computation.*

[180] Hussein Mouzannar, Mesrob I Ohannessian and Nathan Srebro, 'From fair decision making to social equality' (2019) *Proceedings of the Conference on Fairness, Accountability, and Transparency* 359. See also Liu et al (n 179).

most likely to succeed on that basis. The overall aim of reducing inequality is to improve social welfare overall, and so all these authors argue that an assessment of overall welfare needs to be brought into the modelling.

4.2.2. Measurement, Bias and Incentives

We saw in the previous section that welfare economics uses modelling to assess how likely it is that any AI fairness technique will improve social welfare in society. Modelling, together with retrospective analysis of economic data, can also be used to measure whether a particular fairness technique does, or is likely to, achieve its stated objective.

A workshop held at Harvard by the Computing Community Consortium Fairness and Accountability Task Force in 2019[181] reviewed a range of the fairness techniques used by AI developers and identified issues which economists were likely to have the tools to assess. An important question which economists might be able to answer is whether identified bias against particular groups is the result of taste-based discrimination, arising from the individual preferences of the decision-maker, or statistical discrimination which is caused by deciding on the basis of data which are a proxy for relevant characteristics that are not directly identifiable in the data.[182] An example of taste-based discrimination might be preferring to employ men over women, or vice versa. Statistical discrimination might include giving employment preference to candidates who graduated from high-ranking universities, although it is known that socially disadvantaged applicants have a lower chance of admission to those universities.[183] The former is clearly unfair selection, and ought to be controlled for if fairness is desirable. The latter might be considered justifiable if those graduates can be shown to be likely to perform better at the job in question – the unfairness here is not in

[181] David C Parkes and Rakesh V Vohra, 'Algorithmic and economic perspectives on fairness' (2019) *arXiv preprint arXiv:1909.05282*.

[182] ibid, 2:

> [A] decision maker who is unconcerned with the agent's demographics per se, but understands that the demographics are correlated with the fitness of the agent for the task at hand, is said to exhibit statistical discrimination. Given imperfect information about the agent's fitness, the decision maker uses the demographic information to make statistically better decisions.

We should note, though, that this distinction breaks down in the face of stereotyping, taking the most salient characteristic which differentiates one group from another and treating it as a marker of group membership. Pedro Bordalo, Katherine Coffman, Nicola Gennaioli and Andrei Shleifer, 'Stereotypes' (2016) 131 *The Quarterly Journal of Economics* 1753 use as examples red-headed Irishness, or elderly Floridians. In both cases, when investigated statistically, these are inaccurate markers for group membership, but they are particularly salient to observers. The authors' modelling predicts that this will influence beliefs about those group, and this is confirmed through experiment and statistical analysis. 'As the representativeness of tail types increases, beliefs of a group are distorted in the direction of that tail.' (1790).

[183] See n 69.

the selection process for employment, but in the broader social inequality which reduces the chance of university admission.

A subsequent review by Rambachan et al[184] considers how to test whether a fairness algorithm is an accurate predictor of the desired matters, such as suitability for employment. They point out that bias may not be inherent in the algorithm itself, but can sometimes lie in areas for which the algorithm cannot correct:

> [A] raw difference in predictions across groups may arise for three reasons: differences in base rates, differences in measurement error, or differences in estimation error across groups. Intervening at the level of the algorithm may address only the last two components of the disparity by investing in better training data and collecting a better proxy for the outcome of interest. In contrast, the base rate difference is a product of the underlying socioeconomic context itself, not the algorithm. Addressing this component may require more fundamental policy interventions that seek to make changes further upstream to address the differences in observable characteristics.[185]

Identifying what part the base rate difference plays in any remaining bias will both inform others, in particular regulators, how far the algorithm developer has been successful in achieving as much fairness as possible, and also identify the socio-economic structures which need to be addressed, through non-algorithmic means, if societal fairness is to be increased.

Rambachan et al also note that the desired outcome built into some fairness techniques is often set for the convenience of the algorithm designer, rather than as an outcome which is likely to reduce inequality.[186] Economists can test for this failing, identifying whether the designer has favoured convenience over predictive force.[187]

Accurate prediction of the effects of any fairness technique is important because of the possibility that the same decision would have been reached even if the fairness algorithm had not been applied. The Harvard workshop discussed the use of modelling to make *ex ante* predictions about whether an AI fairness technique will achieve its aim of reducing inequality. This is a current research field which is still identifying appropriate economic tools: 'Measuring the effect of an algorithm on an outcome is inherently difficult because decisions made (or influenced) by an algorithm may have happened identically in the absence of the algorithm.'[188] Possibilities discussed include randomised controlled trials, which present real

[184] Ashesh Rambachan, Jon Kleinberg, Jens Ludwig and Sendhil Mullainathan, 'An economic perspective on algorithmic fairness' (2020) 110 *AEA Papers and Proceedings* 91.

[185] ibid, 92.

[186] ibid, 93.

[187] See Ziad Obermeyer, Brian Powers, Christine Vogeli and Sendhil Mullainathan, 'Dissecting racial bias in an algorithm used to manage the health of populations' (2019) 366 *Science* 447; Sendhil Mullainathan and Ziad Obermeyer, 'Does machine learning automate moral hazard and error?' (2017) 107 *American Economic Review* 476.

[188] Parkes and Vohra (n 181) 4.

design difficulties, and the regression discontinuity technique[189] which analyses cases at the border established by the fairness technique.

Roemer and Trannoy[190] have built on Roemer's philosophical work on luck egalitarianism (see section 4.1.3), which argues that inequality of opportunity should be compensated for if it is caused by an individual's circumstances, but that inequalities which arise from differences in the effort expended by individuals should not. This approach to fairness requires some method of quantifying the relative contributions to inequality deriving from effort and circumstances, and the article analyses the possible econometric techniques at some length.[191] However, the authors caution that identifying contribution does not also identify causation – for example, ill-health in children causes inequality in later life, and ill-health of parents correlates with ill-health of children, but this does not necessarily mean that parental ill-health causes childhood ill-health.

> Experiments and quasi-experiments enable one to make causal statements, but experiments can usually only study problems that are much more circumscribed than those that interest researchers in this field. We are trying to understand the whole process by which someone reaches an income level, a health status, or an educational attainment. These processes are dynamic and cover part of the lifespan of an individual, and understanding them fully in a causal way seems out of reach at present.[192]

Measuring what fairness algorithms are actually testing for, and whether they have achieved or will achieve their aims, is clearly useful. However, it is important to remember that the apparent objectivity of economic analysis is far from uncontested. Sociologists might well argue that the assumptions underlying economic theories and models are not objectively true, but that instead they have the meaning which is given to them by the ways in which they are currently used, explained and discussed.[193] Thus the 'reality' of those concepts can shift, and their measurement power also.[194] Meyers and Van Hoyweghen[195] analyse in depth the development of the meaning of actuarial measurements of fairness

[189] For applications of regression discontinuity see Bo Cowgill, 'The impact of algorithms on judicial discretion: Evidence from regression discontinuities' (unpublished manuscript, Columbia Business School, 2018); Richard Berk, 'An impact assessment of machine learning risk forecasts on parole board decisions and recidivism' (2017) 13 *Journal of Experimental Criminology* 193; Ashton Anderson, Daniel Huttenlocher, Jon Kleinberg and Jure Leskovec, 'Steering user behavior with badges' (2013) *Proceedings of the 22nd international conference on World Wide Web* 95; Megan Stevenson and Jennifer Doleac, 'Algorithmic risk assessment tools in the hands of humans' (2019), papers.ssrn.com/sol3/papers.cfm?abstractid=3489440.

[190] John E Roemer and Alain Trannoy, 'Equality of opportunity: Theory and measurement' (2016) 54 *Journal of Economic Literature* 1288.

[191] ibid, 1308–28.

[192] ibid, 1310.

[193] In this impossibly brief explanation I have deliberately avoided technical terminology such as 'construction' and 'performance', originating in the works of Michel Foucault, or Actor-Network Theory's 'enactment'.

[194] See eg Michel Callon, 'An essay on the growing contribution of economic markets to the proliferation of the social' (2007) 24 *Theory, Culture & Society* 139.

[195] Gert Meyers and Ine Van Hoyweghen, 'Enacting Actuarial Fairness in Insurance: From Fair Discrimination to Behaviour-based Fairness' (2018) 27 *Science as Culture* 413, 423:

over time, and investigate its transformation via the introduction of behavioural motor insurance policies using data from monitoring equipment in vehicles. Economic measurements thus play an important but non-definitive role in assessing fairness. They must still be evaluated in combination with other, qualitative, means of fairness assessment as part of the overall fairness package.

One other form of economic measurement is also worthy of consideration: the evaluation of economic incentives to make fair decisions. Regulation for fairness is clearly an important incentive here,[196] but regulation does not (in spite of what some lawmakers seem to believe) guarantee to produce its intended results.[197]

Economics might also help to answer questions about the likely most effective forms of regulation, and particularly whether regulation should focus on the inputs to an AI or the outcomes of its decisions.[198] Input regulation might specify the fairness techniques to be implemented in the AI, and the kinds of data on which it should be trained. Outcome regulation would simply require fair decisions, or reasonable efforts to ensure fair decisions, perhaps expanding somewhat on the criteria against which fairness will be assessed. This choice is evaluated in section 8.1.1 as a preliminary to outlining a regulatory system for AI fairness.

4.2.3. Behavioural Economics

Through experiments using economic games, economists have discovered that human behaviour often fails to accord with those economic theories which assume that human decisions will be based on maximising the individual's utility in terms of their purely material interests: 'Since the 1970s, behavioural economics has been used to investigate actual decision-making situations without imposing expectations about the rationality of behaviour.'[199]

> [T]he technical concept of actuarial fairness has been employed as nonproblematic, rendering it almost invisible in insurance handbooks. It is not deemed necessary to elaborate on the concept ... It is as if actuarial fairness is enacted 'without a doubt.'

[196] Gediminas Adomavicius and Mochen Yang, 'Integrating behavioral, economic, and technical insights to understand and address algorithmic bias: A human-centric perspective' (2022) 13 *ACM Transactions on Management Information Systems (TMIS)* 1; Bo Cowgill and Catherine E Tucker, 'Economics, fairness and algorithmic bias' (2019) in preparation for *Journal of Economic Perspectives*, conference.nber.org/confer/2019/YSAIf19/SSRN-id3361280.pdf.

[197] Chris Reed, 'How to make bad law: lessons from cyberspace' (2010) 73 *Modern Law Review* 903.

[198] Adamovicius and Yang (n 196) 8 propose a holistic approach to this question, assessing the inputs and outcomes together, whereas Cowgill and Tucker (n 196) 2 are firm that outcome regulation is the better approach:

> Effective policy for algorithms should therefore regulate outcomes, not inputs. Regulating inputs – for example, by requiring certain engineering practices or technologies – exhibit the 'command- and-control' approach economists have long opposed in arenas such as environmental policy. Regulations focusing on outcomes exhibit more flexibility, fewer loopholes, greater efficiency and stronger incentives for innovation.

[199] Meyers and Van Hoyweghen (n 195) 416, citing Richard Thaler, *Misbehaving: The Making of Behavioural Economics* (London, Allen Lane, 2015).

Behavioural game theory experiments are used to try to discover the reasons for the differences from what theory predicts, and these seem closely linked to aspects of human psychology. In game experiments, participants are given resources (usually money) and asked to allocate them according to simple rules. Much of the work on fairness has been via the ultimatum game, in which one player is given money and told to offer to share it with the other as the first player sees fit (including offering nothing). If the second player rejects the offer both get nothing, otherwise they share in accordance with the offer. A variation, the dictator game, is where the second player is forced to accept the offer. Both of these find that players behave in apparently economically irrational ways. The offeree in the ultimatum game rejects low offers, even though accepting them would be better than the alternative of nothing. Offerors in the dictator game usually share up to half, even though they would be better off by giving nothing.

By analysing the outcomes of these games, economists develop theories about human emotions and social norms which can be tested via modified games and can demonstrate how far these emotions and social norms are likely to affect human decisions in exchange transactions.

> Along with others, I have tried to pry economists away from narrow assumptions about self-interest. Behavior is driven by a much richer set of values and preferences. [My] analysis assumes that individuals maximize welfare as they conceive it, whether they be selfish, altruistic, loyal, spiteful, or masochistic.[200]

Fairness is one of the most important of these norms and emotions. As examples, players tend to punish other players whose behaviour is perceived as unfair, even though the punishment has an economic cost to the punisher,[201] and where two players are involved in a series of games with each other they tend to converge on cooperative strategies which are based on fairness norms.[202]

At first sight, much of this work does not map directly to the kinds of AI decisions I have been discussing because those AI decisions are imposed on subjects, rather than being decisions as part of an exchange between individuals. However, the work does give us a number of important insights into the nature and strength of fairness norms. Imposed decisions still need to be fair because, as we saw earlier,[203] the legitimate authority of any decision system is derived from its societal acceptance. Fairness in decision-making plays a major part in forming the reactions of those who are affected by the decision.[204]

[200] Gary S Becker, 'Nobel lecture: The economic way of looking at behavior' (1993) 101 *Journal of Political Economy* 385.

[201] Joel Sobel, 'Interdependent preferences and reciprocity' (2005) 42 *Journal of Economic Literature* 392.

[202] Fabian Winter, 'Fairness norms can explain the emergence of specific cooperation norms in the battle of the Prisoner's Dilemma' (2014) 38 *Journal of Mathematical Sociology* 302.

[203] See text to nn 74 and 75.

[204] See Matthew Rabin, 'Incorporating fairness into game theory and economics' (1993) 83 *The American Economic Review* 1281, 1283, arguing that a subject's happiness with a decision (based on its perceived fairness) is an important factor in determining their reaction to it.

Wilson's linguistic analysis[205] suggests that the English language term 'fairness' decomposes in other languages into three distinct elements – equity, justice and fairness – which correspond closely to the distinction I made earlier between outcome fairness (section 3.1), pure procedural fairness (section 3.2) and interactional or 'relational' fairness (section 3.2.1) respectively. He argues that although behavioural economists tend to use the term in its third sense, they are not simply measuring an individual's internal feelings or emotions as influences on that individual's decisions. Rather, they are measuring that individual's private feelings as an element of that individual's interactions with wider society, and thus their findings tell us about the meaning ascribed to fairness by society as well as by the individual: 'Since fairness is not fixed outside regularized human activity, but rather is semantically interpreted within social practice, economic experiments can use repeated interactions to see where people agree as to what is fair.'[206]

There is a clear consensus that the social fairness norm is extremely strong, and plays an important role in determining the individual and collective reactions to decisions. Fairness norms are not an absolute determinant of those reactions, though. Economists have discovered that the effect of fairness norms reduces as the social relationship between those involved becomes more distant,[207] and that other factors such as the desire to maintain social relationships also has an influence and can moderate the effect of fairness norms.[208]

A behavioural game theory analysis may therefore tell us how closely an AI's fairness mechanisms match the social conception of fairness, and also something about the reactions to and acceptance of an AI as decision-maker. It is also possible that behavioural game theory might be useful in improving the design of AI fairness.[209]

4.2.4. Economics and AI Regulation

Successful regulation is not easy to achieve. Two of the main pitfalls are misidentifying the problem which regulation is intended to solve, and adopting regulatory solutions which fail to achieve their desired ends.[210] An understanding of the work which economists have done on fairness can assist here.

[205] Wilson (n 11).

[206] ibid, 424. See also Greenberg (n 21) on the social emotion of fairness.

[207] See Sobel (n 2010) 419, suggesting that in anonymous markets there are lower incentives for reciprocal behaviour, such as fairness. But see contra Philippe van Basshuysen, 'Markets, market algorithms, and algorithmic bias' (2023) *Journal of Economic Methodology* 310, arguing that algorithmic market design needs to accommodate fairness norms.

[208] Catherine C Eckel and Philip J Grossman, 'The relative price of fairness: Gender differences in a punishment game' (1996) 30 *Journal of Economic Behavior & Organization* 143.

[209] Vincent Conitzer, Walter Sinnott-Armstrong, Jana Schaich Borg, Yuan Deng and Max Kramer, 'Moral decision making frameworks for artificial intelligence' (2017) 31 *Proceedings of the AAAI Conference on Artificial Intelligence* 4831.

[210] Reed (n 197).

First, that work provides empirical support via data analysis, modelling and experiment for many of the elements of the social and psychological conceptions of fairness examined in chapter three. This confirms that any regulation of AI development and use ought to be based on three fundamental principles:

- Fairness in individual decision-making should be aimed for, and any system of AI decision-making should be measurably fair in aggregate even if some individual decisions fall short.
- A focus purely on individual decisions is not enough. A fair AI will enhance fairness in society overall, in addition to achieving individual fairness.
- Fairness is not the sole criterion for judging the acceptability of an AI decision-maker. Fairness should not be achieved at the expense of other important considerations, including efficiency and the maintenance of social harmony.

However, there is no doubt that where fairness regulation is appropriate, it needs to demand an acceptable level of fairness from AI decision-makers. The perception of fairness by individuals and society, both of individual decisions on a cumulative basis and the overall fairness of the decision-system, will strongly influence the acceptance of those decisions and thus the legitimate authority of the system.

Second, there is a range of econometric tools which are potentially useful for objective assessment of AI fairness techniques. Regulation will wish to ensure that AI developers adopt fairness techniques which are actually successful in reducing inequality and bias – mere faith or aspiration that they will do so is insufficient. Economic analysis can help in three ways:

- By identifying the contribution which existing inequality of opportunity makes to the social situation, thus telling developers where to focus their algorithms.
- By identifying how far a particular fairness algorithm addresses the actual (rather than presumed) causes of inequality, and how far it is actually likely to succeed in addressing that inequality.
- By helping to identify the most likely effective forms of regulatory intervention based on analysis of the incentives that the regulation generates.

There is consistent agreement that more complete and accurate data would enable economists to make more accurate assessments, but unfortunately this clashes with important legal principles such as privacy and data protection (see section 5.3.2). Thus economics cannot provide definitive answers, even within its own limited domain of application, but the assessments it *can* provide will still be of use to lawmakers.

One important caveat is that economics is not necessarily useful for discovering the *causes* of inequality, though it may be able to eliminate some potential causes and show that others have only a partial influence on inequality. Certainly, any decision about what action, if anything, ought to be taken to remedy an identified inequality is driven by the philosophical approach adopted by the social engineer, such as a lawmaker. A policy and supporting regulation based on Roemer's idea of remedying inequality arising from circumstances, but not effort, would look very different from one which attempts to implement Fleurbay's concept of outcome equality.

Finally it is worth noting that there is some support in the economic literature for the idea, already discussed,[211] that AI developers should be required to justify the normative choices they make, and further explain the assumptions underlying their fairness measures and the testing of outcomes they have undertaken. Kasy and Abebe[212] propose a detailed methodology using econometric techniques for fairness testing, while Lee et al set out a justification process using Key Ethics Indicators.[213] Li and Chignell[214] take an existing engineering method for failure mode assessment and modify it to test for algorithmic fairness. This disciplinary convergence on justifying fairness choices for AI reinforces the idea that requiring such justifications might be an important regulatory tool.

4.3. Rules and Choices

One of the main reasons why the AI fairness community has been so attracted to the work of philosophers is their desire for a clear model or rule set which, if followed, will enable them to declare that their AIs are fair. Sadly, as we have seen, there is no such model or rule set to be found. Ethicists set the goal, in this case fairness, but ethics does not explain how that goal is to be achieved. Philosophy comes closer to setting out rules, but there is substantial disagreement among philosophers as to which are the correct rules. There is a consensus that fairness requires equality of treatment, but whether that should be through equality of opportunity or equality of social welfare is hotly contested, as also are the mechanisms proposed to equalise opportunity or welfare.

What we find therefore is that there are numerous suggested routes to achieve fairness, each of which makes different assumptions and prioritises some characteristics of the decision subject over others. Any one of them is likely to achieve a fair result if viewed by those who share the same assumptions, but will be seen as less fair (or even unfair) by those whose assumptions are different. This means that AI producers need to make choices about how they plan to achieve fairness, which also means they need to understand the underlying assumptions of their chosen fairness method and what its consequences are likely to be.

[211] See sections 3.1.3, 3.2.1, 3.2.2, 4.1.1 and 4.1.6.

[212] Kasy and Abebe (n 44) 583.

[213] Lee et al (n 31) 539–42.

[214] Jamy Li and Mark Chignell, 'FMEA-AI: AI fairness impact assessment using failure mode and effects analysis' (2022) *AI and Ethics* 1.

To some extent, the work of economists has helped here by clarifying what those choices maximise for, and how far they achieve their intended results. However, economic analysis requires data, and much of the data needed for a proper analysis is either subjective, such as individual preferences, or simply missing. The need for data also means that economic analysis of a particular fairness technique, whether implemented in regulation or built into an AI, can only be carried out after enough time has passed for that data to become available.

Law and regulation, in contrast, do set out some fairness rules in advance, and compliance with these is required if there is a legal obligation to achieve fairness. However, as we will see in the next chapter, those rules only cover a subset of the fairness issues which we have identified so far. This is because they were designed with human decision-makers in mind, and thus assume that the decision-maker understands society's collective conception of fairness. We need to ask what AI producers must do to comply with those rules, and identify which aspects of fairness they might deal with successfully if implemented in an AI decision system. However, throughout we also need to remember that rule compliance only goes part of the way to achieving fairness.

5

Law's Conception of Fairness

Law adopts two complementary approaches to the issue of fairness. The first sets out rules for fair procedures which should be followed when making a decision, while the second focuses on whether the decision outcome is a fair one.

Law's fairness questions only arise if the decision-maker has a legal duty to act fairly, as there is no generally applicable legal obligation to be fair. As a broad generalisation, those who are performing some public function normally have a duty to make fair decisions.[215] For other decisions where private rights are affected, a duty of fairness needs to be found in statute, regulation or the common law. As an example, the Equality Act 2010 imposes a duty not to engage in unlawful discrimination on a wide range of public and private sector actors.

From an AI producer's perspective, it might be a sensible heuristic to build in the elements of procedural fairness so far as possible, even where there might not be a duty to adopt fair procedures. This will allow for the possibility that a court finds an obligation of procedural fairness to subsist and, perhaps more importantly, could help to achieve public acceptance of the AI's use as explained in section 3.2.

5.1. Procedural Fairness

Procedural fairness is explained in the greatest detail in that branch of the law which controls the actions of public bodies. In the UK that is described as administrative law, but in jurisdictions such as the US which have fairness obligations in their constitutions it will be part of that country's constitutional law.

In the UK, the overriding principle is that decisions of those exercising public functions should be made in a fair way, as a matter of natural justice – 'justice is intuitively understood to require a procedure which pays due respect to persons whose rights are significantly affected'.[216] This principle breaks down into a number of elements which are often described as the 'rules' of natural justice in the literature, but it is clear from the cases that how far each rule needs to be followed by a particular decision-maker depends on the nature and context of the decision.

[215] In the UK this extends to private bodies which the courts perceive as exercising public functions – see *R v Panel on Takeovers and Mergers ex p Datafin plc* [1987] QB 815.
[216] *Osborn v The Parole Board* [2013] UKSC 61 [67]–[72] per Lord Reed.

The principles of natural justice are often expressed as two Latin maxims:

- *Nemo judex in re sua*, or 'No man a judge in his own cause'. Initially this prohibited decisions from being made by a person who might benefit personally from a particular outcome, for the obvious reason that this might influence their decision. It has been extended to cover prejudgement or actual bias, whether or not this benefits the decision-maker, and also to decisions made in circumstances which might lead outsiders to suspect prejudgement or bias.
- *Audi alteram partem*, or 'Hear the other side'. Where there is a formal hearing, each of the parties affected must be allowed to state the case for a decision in their favour. This principle also extends to non-adversarial decision-making without a formal hearing, such as land use planning decisions where not only the applicant but also others who believe themselves to be affected are entitled to make representations about the proper outcome.

A decision which is reviewed under UK administrative law and found to have been made in breach of these principles is normally void, though the reviewing court may require the decision-maker to reconsider the case in accordance with natural justice.

5.1.1. Avoidance of Bias

To comply with this principle a decision-maker must neither be biased nor make the decision in circumstances which might lead to a reasonable suspicion of bias. There are three main ways in which such a suspicion of bias might be generated.

First, if the decision-maker has an interest which is likely to be advanced by the outcome of the decision. The most obvious example is where the decision-maker owns property whose value will be increased by the decision. In *Dimes v Grand Junction Canal*,[217] where a legal dispute between the canal company and a landowner was decided in favour of the company, it was later discovered that the judge owned shares in that company. The reviewing court stated that although there was no suggestion that the judge had actually been influenced by this fact, nonetheless the case should be re-heard by a different judge who had no financial interest in the outcome.

Second, if there is a prior relationship between one of the parties and the decision-maker that can give rise to a suspicion of bias. In *Lawal v Northern Spirit Ltd*[218] one of the barristers in an employment case was also a part-time judge. The employment appeal tribunal in the case consisted of a judge and two lay members, and the barrister had previously sat as a judge with one of those lay members when deciding a different case. The House of Lords held that this

[217] (1852) 3 HLC 759. See also *R v Sussex Justices ex p McCarthy* [1924] 1 KB 256.
[218] [2003] UKHL 35.

previous relationship created the risk that the lay member would, perhaps unconsciously, give extra weight to the barrister's arguments, and thus the circumstances might lead an observer to suspect bias.

Finally, a decision can be voided if there is a reasonable suspicion that the decision-maker has prejudged the case. Prejudgement is not bias per se, but it raises the same issue – a decision-maker must form their view of the correct outcome on the merits of the case alone, and prejudgement raises a suspicion that the merits were not properly considered, just as much as bias does. Evidence for the risk of prejudgement might come from previous statements of the decision-maker,[219] or because the decision-maker had been involved in making some previous decision which is relevant to the matter.[220]

Difficulties in applying this principle of natural justice arise where the decision-maker is not an independent judge, but instead is some policy-maker who is also charged with making decisions. It is inevitable in such cases that the existence of policy will, to some extent, pre-determine or engender bias in favour of a particular outcome. The normal remedy is to substitute a different decision-maker, but as *Wade and Forsyth's Administrative Law* points out:

> [T]here are many cases where no substitution is possible, since no one else is empowered to act. Natural justice then has to give way to necessity; for otherwise there is no means of deciding and the machinery of justice or administration will break down.[221]

In these cases the courts have tended to confirm decisions so long as the decision-maker still appears to be able to decide on the merits, recognising that is inevitable that non-judicial decision-makers will have policy positions which influence decision outcomes.[222]

An important question for AI producers is whether procedural fairness demands the absence of bias or prejudgement, in which case it might be possible to demonstrate empirically that the AI was procedurally fair, or whether the mere reasonable suspicion of bias or prejudgement is enough to void a decision. It is now clear, in the light of the jurisprudence under Article 6(1) of the European Convention on Human Rights, that the answer is 'both'.[223] Thus the UK courts now focus primarily on suspicion of bias, applying the test of a reasonable, external observer.[224] AI developers cannot test empirically for such suspicion, which requires a qualitative evaluation of the AI's decision-making by an external observer.

[219] *National Assembly for Wales v Condron* [2006] EWCA Civ 1573.

[220] See eg *Cooper v Wilson* [1937] 2 KB 309, an appeal against the dismissal of a police officer by the chief constable. In that case, the chief constable was present during the deliberations of the appeal body, and this gave rise to suspicion that his prior judgement in the case might have influenced the appeal body's decision.

[221] William Wade, Christopher Forsyth and Julian Gosh, *Wade and Forsyth's Administrative Law*, 12th edn (Oxford, Oxford University Press, 2022) 379.

[222] See eg *Re Manchester (Ringway Airport) Compulsory Purchase Order* (1935) 153 LT 219.

[223] *Piersack v Belgium*, A 53 (1982) 5 EHRR 169.

[224] *Porter v Magill* [2001] UKHL 67; *Lawal v Northern Spirit Ltd* [2003] UKHL 35.

Although the AI community has focused extensively on identifying and countering *actual* bias (see section 7.4), the concentration has been on biased decision outcomes, not procedure. At first sight it might seem obvious that if an AI can be shown not to make biased decisions, no issues of procedural bias arise. The AI has no financial or other interest in the outcome, nor does it have any desire to validate its previous decisions by deciding later cases the same way (though it is of course highly likely to do so). However, this still leaves two possibilities for procedural unfairness:

- Empirical testing for bias is not objective. The criteria used for data labelling may fail to eliminate the possibility that the AI learns human biases from its training data.[225] More important, all empirical methods of testing AI for bias rest on underlying assumptions about what kinds of equality are desirable (see section 4.1) and what groups should be tested for equality – these decisions by AI developers are not by any means objective. It is conceivable that, viewed holistically, a decision-making system incorporating AI might still be found to be a biased or prejudging decision-maker.
- Reasonable suspicion of bias is a subjective state of the observer, not an empirically determinable matter. Observers of an AI decision system can only form a view on the basis of the information available to them, which suggests that providing explanations of the AI producer's approach to identifying and eliminating unintended bias,[226] and also explanations which justify individual decisions, might be useful ways to reduce such suspicions and thus achieve procedural fairness.

5.1.2. Fair 'Hearing'

Procedural fairness demands that a person whose rights or interests will be affected by a decision should be able to make representations to the decision-maker. Formal hearings are only used for a minority of decisions, so this should be seen more broadly as a requirement for a fair process under which each person whose rights or interests are affected has an opportunity to put the case in favour of their preferred outcome. In adversarial proceedings this should include the

[225] An example might be training data which contains no labels for race but does include postcodes. Racial geographic clustering is a known phenomenon, and thus postcode data can be an unintentional proxy for race. For an overview of the issues created by data labelling, see Daochen Zha, Zaid Pervaiz Bhat, Kwei-Herng Lai, Fan Yang, Zhimeng Jiang, Shaochen Zhong and Xia Hu, 'Data-centric artificial intelligence: A survey' (2023) *arXiv preprint arXiv:2303.10158*, 10–12.

[226] Every AI decision system has some element of intentional bias because it is deciding resource allocation on the basis of merit criteria set by the AI user. The simplest example is an employer seeking to employ the most suitable job applicant. Here the system is biased against less suitable candidates, but not unfairly so. Other biases, which were not explicitly built into the system as merit criteria, are most likely to be where fairness issues potentially arise.

opportunity to make representations to counter the opposing case, which thus requires the person affected to be given knowledge of it.[227]

This right to present a case arises whenever the issue is of a 'judicial' nature, a broad concept which extends beyond purely public decision-making.[228] Thus the courts have decided that decisions to expel a member from a club[229] or from a mutual insurance society[230] engaged the fair hearing principle on the basis that this was an implied term of the contract between members. The courts do not seem yet to have been required to clarify the borderline between judicial and non-judicial decisions, but a useful distinction might be whether the decision has the potential to reduce or extinguish the rights or privileges[231] of an individual. Thus, as an example, a decision whether to employ someone is not a judicial decision because they currently have no right to be employed. However, a decision to terminate their employment does have a judicial character and therefore normally requires that the employee should have the chance to respond before the decision is made.[232]

Whether the courts might find a fair hearing duty for a decision which allocates opportunity, rather than taking away rights, is not easily predictable. The courts have consistently made it clear that the rules of natural justice are not granular, but rather that they are all examples of a more general overriding principle: 'As has frequently been said, and there is no harm in repeating it, all that the rules of natural justice mean is that the proceedings must be conducted in a way which is fair … fair in all the circumstances.'[233]

If we take the example of an AI which allocates scarce medical treatment under the UK's National Health Service, it is arguable that the AI is undertaking a public function by allocating public resources because healthcare is paid for from public funds. Even though no patient has a right to such treatment, it might be tempting for a court to decide that the AI must still use decision-making procedures which are fair. If the AI is calculating the potential benefit from the treatment via patient information supplied by medics, rather than directly by the patient, the fair hearing obligation would mean the AI needed to take into account all the relevant information which the patient had supplied to the medic. Reliance on the medic to supply accurate information would seem a fair process, but there is a clear risk that the medic might have failed to ask the patient for relevant information, or failed to record what was supplied. Making the decision on

[227] *Thirukumar, R (on the application of) v Secretary of State for the Home Department* [1988] EWHC 1 (QB).

[228] Wade et al (n 221) 396–97 discussing *Ridge v Baldwin* [1964] AC 40.

[229] *Dawkins v Antrobus* (1881) 17 Ch D 615.

[230] *Wood v Woad* (1874) LR 9 Ex 190.

[231] This is the phrase used multiple times in *Ridge v Baldwin* (n 228), the case which is thought to have established the modern UK law on judicial review. See also *R v Commission for Racial Equality ex p Hillingdon LBC* [1982] AC 779, 787 per Lord Diplock.

[232] *Polkey v AE Dayton Services Ltd* [1987] UKHL 8.

[233] *R v Commission for Racial Equality ex parte Cottrell and Rothon* [1980] 1 WLR 1580, 1586 per Lord Lane CJ.

only partial data presents a clear risk of unfairness to the patient. The fair hearing duty might thus require the AI to prompt the medic for potentially missing information, leaving it to the medic to obtain that information from the patient.

An interesting question for producers of an AI decision system which raises procedural fairness issues is whether they should build in some review or appeal process. AIs do not hold hearings or receive representations from affected individuals, except to the extent that information provided by those individuals might be inputs to the AI, and thus AI producers have little opportunity to meet the fair hearing obligation. Might an appeal process resolve this issue?

The UK courts have decided that the rules of natural justice do not automatically require an opportunity of appeal to be offered,[234] and are in some disagreement about whether an appeal which meets the rules of natural justice can substitute for an initial decision which does not.[235] Bearing in mind that in the case of an AI decision system there is no possibility of reconsideration by the AI in a way which might constitute a fresh hearing, I take the view that a process of human review or appeal should be seen by the courts as part of the original decision-system as a whole, so that they need to assess whether the whole process meets the rules of natural justice and not just the original decision which the AI made. This accords with Megarry J's analysis in *Leary v National Union of Vehicle Builders*[236] of the Canadian decision in *King v University of Saskatchewan*.[237] In that case the University decided to refuse the award of a degree, following a hearing by the Senate by way of appeal from the original decision of the academic department concerned. The university's constitution conferred the power to award degrees on the Senate, and Megarry J thought that compliance with natural justice by way of a fair hearing before the Senate was sufficient, even though it was not complied with for the original decision.

To meet the fair hearing requirement, decision-makers must also give effect to any legitimate expectations which they have raised in the minds of affected individuals. In particular, if the decision-maker has published its decision-making criteria it must decide in accordance with them – any additional, secret considerations will be a breach of the rules of natural justice.[238] This will put substantial importance on any statement which the AI developer publishes explaining the fairness considerations which have been designed into the AI. Inaccuracy here could render all the AI's decisions void. Fortunately, as we saw in section 4.1.6,[239] the process of formulating explanations increases the chance that the

[234] *Ward v Bradford Corporation* [1971] 7 WLUK 39.

[235] Contrast *Ridge v Baldwin* (n 228) 76 per Lord Reid (stating that a speedy reconsideration can cure the defect in the original decision) with Megarry J's judgment in *Leary v National Union of Vehicle Builders* [1971] Ch 34, holding that only a re-hearing by the original decision-maker will do.

[236] ibid.

[237] (1969) 6 DLR (3d) 120.

[238] *R v Secretary of State for the Home Department, ex p Khan* [1984] EWCA Civ 8; *Council of Civil Service Unions v Minister for the Civil Service* [1985] AC 374.

[239] See Peters (n 167).

decision-making will follow that explanation and thus comply with the rules of natural justice.

Finally, it is worth noting that the fair hearing duty has been expanded by the UK courts to encompass a number of other matters which include:

- a requirement for decisions to be based on evidence which supports the outcome; and
- a requirement in some cases to justify the decision by giving reasons for it.[240]

In my view these are more appropriately seen as aspects of outcome fairness, and so are discussed in the next section.

5.2. Outcome Fairness

It is reasonably accurate to say that the law, in the UK at least, does not impose any general obligation to treat others fairly when making decisions. Such an obligation can be imposed by law or regulation in particular relationships, or might be implied into contractual terms if the context of the contract suggests that this is what the parties in fact agreed, but otherwise a decision can be as capricious as the decision-maker likes. An employer can refuse to employ someone just because they have taken an irrational dislike to them, so long as they are not also discriminating on some unlawful ground.

However, if the decision is subject to judicial review under the rules of natural justice, then there appear to be some general outcome fairness duties which must be complied with. The first of these is a requirement for decisions to be based on evidence which logically supports the outcome. In *R v Deputy Industrial Injuries Commissioner ex p Moore*[241] the Commissioner's decision to deny an industrial injury claim was based in part on medical evidence from previous cases about similar types of injury. The reviewing court held that this evidence had no evidential connection to the specific claimant's injury, so that the Commissioner's reliance on it breached the rules of natural justice.[242] In *Mahon v Air New Zealand Ltd*[243] the judge who led a Royal Commission investigation into an air crash made findings of fact which were not supported by the evidence. The Privy Council held that this infringed the rules of natural justice because

[240] For an analysis of how these rules might apply in the criminal justice field, examining related issues to those discussed here, see Marion Oswald, 'Algorithm-assisted decision-making in the public sector: framing the issues using administrative law rules governing discretionary power' (2018) *Philosophical Transactions of the Royal Society A*: 376(2128):20170359.

[241] [1965] 1 QB 456.

[242] ibid, 476 per Willmer LJ: '[T]he procedure must be in accordance with natural justice. This involves that any information on which the commissioner acts, whatever its source, must be at least of some probative value.'

[243] [1984] AC 808.

any evidence relied on needs to have 'some probative value', which means it must be based upon 'some material that tends logically to show the existence of facts consistent with the finding and that the reasoning, if it be disclosed, is not logically self-contradictory.'[244] A decision which is not based on evidence is an unfair decision, and suspicion alone is not enough without grounds to support it.[245]

Second, in some cases the courts have held that natural justice requires decision-makers to give an explanation of the reasons for their decision. This clearly links to the requirement that the decision should be supported by evidence, as the purpose of the reasons is to convince their recipients that the decision was properly arrived at.

> [I]f all other known facts and circumstances appear to point overwhelmingly in favour of a different decision, the decision maker who has given no reasons cannot complain if the court draws the inference that he has no rational reason for his decision.[246]

This requirement of natural justice is also linked to the principle that fairness often[247] requires that the person affected be given the chance to make representations, and those representations cannot be meaningful without knowing the reasons for the decision.[248] Meaningful reasons 'should contain sufficient [information] to enable the reader to know what conclusions have been reached on the principal important issues', identifying the reasoning behind the decision and explaining which factors justified that decision.[249]

Poole[250] has suggested that the rules of natural justice are beginning to work in tandem with the UK Human Rights Act 1998 to develop a requirement that all decisions which are subject to judicial review should meet the test of proportionality. Proportionality requires that any decision which affects an individual's fundamental rights should attempt to strike a fair balance between those

[244] ibid, 820–21.

[245] *Khawaja v Secretary of State for the Home Department* [1983] UKHL 8; *Bateman & Bateman, R (on the application of) v Legal Services Commission* [2001] EWHC Admin 682.

[246] *R v Trade Secretary ex parte Lonrho plc* [1989] 1 WLR 525, 540. See also *Belfast City Council v Miss Behavin' Limited* [2007] UKHL 19 [37] per Baroness Hale:

> The views of the local authority are bound to carry less weight where the local authority has made no attempt to address that question. Had the Belfast City Council expressly set itself the task of balancing the rights of individuals to sell and buy pornographic literature and images against the interests of the wider community, a court would find it hard to upset the balance which the local authority had struck.

[247] It has been suggested that this applies, with the result that reasons need to be given, where a decision-maker is acting judicially rather than domestically, or where there is no right of appeal, or where the decision-maker has raised a legitimate expectation that reasons should be given – *R v Civil Service Appeal Board, ex parte Cunningham* [1992] ICR 816, 826–28.

[248] *R v Home Secretary ex p Doody* [1994] 1 AC 531, 560. See also *Kanda v Government of Malaya* [1962] AC 322, 337.

[249] *Bateman* (n 245) [116]–[117]. See also *South Bucks District Council v. Porter (No 2)* [2004] UKHL 33 [26].

[250] Thomas Poole, 'The reformation of English administrative law' (2009) 68 *Cambridge Law Journal* 142.

rights and the other rights and interests at issue. It requires justification of the decision,[251] and thus not merely an explanation for it, but also an explanation which identifies the fairness of the balance. Proportionality is so closely related to fairness[252] that this trend might develop into a general obligation to make fair decisions, though it is clear that UK administrative law has not yet reached that position.

In decision-making which falls outside the scope of administrative law, however, outcome fairness obligations will generally only subsist if they are imposed by law or regulation. The next section examines three examples, in an attempt to understand what law and regulation might require to achieve outcome fairness.

5.3. Express Outcome Fairness Obligations

5.3.1. Avoiding Unlawful Discrimination

We noted in chapter two that most states have some conception of unlawful discrimination. Usually the prohibited grounds for discrimination are set out by legislation, such as the Equality Act 2010, section 4, which prohibits discrimination on the grounds of age, disability, gender reassignment, marriage and civil partnership, pregnancy and maternity, race, religion or belief, sex, and sexual orientation. If an individual with one or more of these characteristics is, directly or indirectly, treated less favourably than another without the characteristic[253] then that individual will have some remedy against the discriminator.[254]

The law requires that individuals who are members of these protected groups should be treated equally with other members of society. The law seeks equality of treatment, not merely the protection of the disadvantaged, so it is equally unlawful to discriminate against a member of a majority or advantaged group, such as (in the UK) a white man.[255]

Equal treatment is a core element of all the conceptions of fairness we have reviewed in previous chapters. It is therefore unsurprising that AI developers have focused heavily on Rawls' philosophical concept of equality of opportunity (see section 4.1) by searching for techniques to identify and eliminate bias against groups in their AI's decisions. Those groups have tended, again unsurprisingly, to be identified by reference to whichever national equality law is most relevant to the developer.

[251] Kai Möller, 'Justifying the culture of justification' (2019) 17 *International Journal of Constitutional Law* 1078.

[252] See the discussion of data protection fairness and proportionality in section 5.3.2 below.

[253] Equality Act 2010, s 13(1).

[254] In the UK there is a complex range of potential remedies depending on the relationship in which the unlawful discrimination takes place – ibid, ss 113–41.

[255] See eg *Bayfield v Wunderman Thompson (UK) Ltd* [2021] 7 WLUK 578 5 Jul 2021.

Unfortunately national laws do not agree on a single list of unlawful discrimination grounds, as these are culturally and socially determined by their respective societies.[256] Thus the list of prohibited grounds in section 9(3) of the South African Constitution, implemented in South African law via the Promotion of Equality and Prevention of Unfair Discrimination Act 2000 extends to: 'race, gender, sex, pregnancy, marital status, ethnic or social origin, colour, sexual orientation, age, disability, religion, conscience, belief, culture, language and birth.'

This differs in several respects from the UK list. Additionally, the UK list is closed, whereas the South African list is open, allowing the courts to identify additional grounds of unfair discrimination. To add further complication, section 9(2) of the South African Constitution specifically permits affirmative action which is 'designed to protect or advance persons, or categories of persons, disadvantaged by unfair discrimination', and by definition such action will inevitably, but legitimately,[257] discriminate against non-disadvantaged persons.

National laws also differ in how unlawful discrimination is defined. In the UK, the test is one of less favourable treatment subject only to a defence of an objective justification for that treatment. In South Africa, by contrast, the law imposes a duty not to 'unfairly discriminate',[258] and sets out a complex set of factors which a court must consider in deciding whether discrimination was unfair.[259] The US law on this topic is distributed among numerous federal and state laws, but many of the fundamental elements of anti-discrimination law have been derived from interpretations of the 14th and 5th Amendments to the US Constitution in cases seeking the annulment of laws which are alleged to be discriminatory. In those cases the courts have held consistently that the relevant question is whether the law had discriminatory *intent*; its discriminatory effect is irrelevant unless that effect combined with other evidence amounts to convincing evidence of the necessary intent.[260]

[256] For an interesting discussion of the reasons why law might protect sex and race, but not height or hair colour, see Michael P Foran, 'Grounding Unlawful Discrimination' (2022) 28 *Legal Theory* 3.

[257] See *Minister of Finance and Other v Van Heerden* (CCT 63/03) [2004] ZACC 3.

[258] Promotion of Equality and Prevention of Unfair Discrimination Act 2000, s 6.

[259] ibid, s 14. For detailed analysis see Grete S Vogt, 'Non-discrimination on the grounds of race in South Africa – with special reference to the Promotion of Equality and Prevention of Unfair Discrimination Act' (2001) 45 *Journal of African Law* 196.

[260] See eg *Washington v Davis*, 426 US 229 (1976), seeking annulment of a state law introducing an employment test which was disproportionately failed by black applicants. The court stated at 239 that:

> The central purpose of the Equal Protection Clause of the Fourteenth Amendment is the prevention of official conduct discriminating on the basis of race … But our cases have not embraced the proposition that a law or other official act, without regard to whether it reflects a racially discriminatory purpose, is unconstitutional solely because it has a racially disproportionate impact.

The detailed analysis at 239–48 explains that the court needed to be convinced that the motive behind the law was to achieve racial discrimination, and that the claimant had not succeeded in doing so.

This does not mean that the stated intent of legislation or a decision is definitive as to intent. 'Despite deploying the mens rea-inflected language of "intent," courts in constitutional antidiscrimination cases necessarily infer wrongful discrimination from the circumstances.' – W Kerrel Murray, 'Discriminatory Taint' (2021) 135 *Harvard Law Review* 1190, 1193. As an example, adopting an

A further complication which is particularly troubling for AI developers is that there is a tendency for national laws to encompass indirect, as well as direct, discrimination. In the UK, the Equality Act 2010 defines indirect discrimination as discrimination on the basis of a non-protected characteristic which has the effect that persons having a protected characteristic are treated less favourably than others.[261] An example from the employment field might be a requirement to work flexibly, on weekends and evenings. Women are, on average, demonstrably less likely to meet such a criterion because of childcare responsibilities, and thus are indirectly discriminated against.[262]

This is a problem for AI developers because the machine learning process identifies combinations of characteristics which, based on previous decisions, suggest a particular outcome. It is feasible to cleanse the training data of characteristics on the ground of which it is expressly unlawful to discriminate. This cannot solve the problem of indirect discrimination though, because other characteristics may be proxies for those characteristics – as an example, gaps in an employment record may be due to maternity leave. An AI which makes decisions on the basis of proxies which do not adequately predict outcomes is at best problematic, and may even be inherently discriminatory.[263]

Anti-discrimination law is complex and nuanced. It is local and culturally determined. It can apply even where there is no discriminatory intent, and in spite of the steps taken by AI developers to avoid discrimination. Some, but not all, of the these difficulties have been recognised by the AI community, and are explored further in chapter seven.

5.3.2. Data Protection and Outcome Fairness

Fairness is a core concept of EU data protection law,[264] but no attempt is made to define fairness in the legislation. Its legal meaning for data protection cases therefore has to be induced from the statements of data protection regulators and the decisions of the courts, informed by the views of academic commentators.

Article 8 of the EU Charter of Fundamental Rights and Freedoms establishes the right to data protection and provides that personal data must be processed 'fairly for specified purposes and on the basis of the consent of the person concerned or some other legitimate basis laid down by law.' The substantive EU

unusual procedure might, in combination with the discriminatory effect, lead a court to find that the underlying intent was in fact discriminatory – Joseph Landau, 'Process Scrutiny' (2019) 119 *Columbia Law Review* 2147.

[261] Equality Act 2010, s 19.

[262] *Dobson v North Cumbria Integrated Care NHS Foundation Trust* [2021] 6 WLUK 306.

[263] See Solon Barocas and Andrew D Selbst, 'Big data's disparate impact' (2016) 104 *California Law Review* 671, 706–09.

[264] European Data Protection Supervisor, 'Opinion on Coherent Enforcement of Fundamental Rights in the Age of Big Data' (EDPS 2016) Opinion 8/2016, 8.

law is found in the General Data Protection Regulation (GDPR),[265] Article 5(1)(a) of which provides that personal data must be 'processed lawfully, fairly and in a transparent manner'. The UK Data Protection Act 2018, at least for the present,[266] mirrors the EU law.

Clifford and Ausloos[267] analyse this fairness concept in depth and propose that it has two distinct roles to play: first as an explicit prerequisite to lawful processing; and subsequently as an implicit obligation when undertaking further processing:

> The express mentioning of fairness and the positioning of fairness, lawfulness, and transparency as 'core' principles is manifested in the recognition of both an explicit and implicit role for fairness. Explicitly in the requirement for fair and transparent processing (ie information requirements and thus the application of the inherent transparency principle) and implicitly via the recognition of the need for a 'fair balance' regarding competing fundamental rights and interests.[268]

5.3.2.1. Procedural Data Protection Fairness

The first role is at the time of collection of personal data, and requires fairness and transparency towards the data subject as part of the controller's justification of a lawful basis for processing under Article 6 of the GDPR. This interpretation is supported by Malgieri's linguistic analysis of the different GDPR texts,[269] which finds that some of them use a term which in English might be expressed as 'loyalty' when referring to the conditions for establishing the lawfulness of processing. He suggests that the issue here is about rectifying the power imbalance between controller and subject. In those languages, loyalty implies that the controller should not use the power imbalance in its favour to privilege its own interests over those of the data subject, but should instead seek an appropriate balance between them which respects both sets of interests.

Because fairness at the time of collection (including fairness in the degree of transparency) is required before any processing takes place, I suggest this first element might usefully be viewed as a procedural fairness requirement, rendering subsequent processing unlawful if it is not followed. The information which Article 13 requires a controller to provide to the data subject, together with any

[265] Regulation (EU) 2016/679 of the European Parliament and of the Council of 27 April 2016 on the protection of natural persons with regard to the processing of personal data and on the free movement of such data, and repealing Directive 95/46/EC (General Data Protection Regulation), OJ L119/1, 4 May 2016.

[266] At the time of writing the Data Protection and Digital Information (No 2) Bill 2023 is under consideration by the UK Parliament. It will retain the obligation to collect and process data fairly, but some of its provisions might change the ways in which fairness is interpreted by the courts, particularly in relation to consent by the data subject.

[267] Damian Clifford and Jef Ausloos, 'Data protection and the role of fairness' (2018) 37 *Yearbook of European Law* 130.

[268] ibid, 138.

[269] Gianclaudio Malgieri, 'The concept of fairness in the GDPR: a linguistic and contextual interpretation' 2020 *Proceedings of the 2020 Conference on fairness, accountability, and transparency* 154, 161–63.

other information which is necessary to achieve fairness,[270] must be provided 'at the time when personal data are obtained', and is thus clearly a prerequisite to any further processing. Implicitly, further transparency is a procedural requirement if the nature of the processing changes, such as by processing the data for a new purpose or transferring or disclosing data to a new controller.[271] Transparency is not required for its own sake, but to ensure that the data subject is treated fairly. For example, where consent is secured as the lawful basis for processing[272] that consent will have been obtained unfairly if the data subject does not know how the data are to be used,[273] or if its planned use does not match the data subject's reasonable expectations.[274] Similarly, a state of ignorance about how data is being processed also unfairly prevents any effective exercise of the rights to object to processing[275] and not to be subject to automated decision-making.[276]

It is worth noting here that some of the other procedural requirements for data collection might generate fairness risks, particularly the requirement in Article 5(1)(c) GDPR for data minimisation which requires that the collection and processing of personal data should be 'limited to what is necessary in relation to the purposes for which they are processed'. Experiments on AI recommender systems have identified that using only the most recent user data does not materially reduce the accuracy of the system overall, but that the accuracy of recommendations to individuals can be greatly reduced unless a longer time span of their data is included.[277]

[270] See recitals 58, 60–62.

[271] See GDPR, recital 60: 'The controller should provide the data subject with any further information necessary to ensure fair and transparent processing taking into account the specific circumstances and context in which the personal data are processed.'

[272] GDPR, Arts 6(1)(a), 7.

[273] GDPR, recital 42. See also European Data Protection Board, Guidelines 05/2020 on consent under Regulation 2016/679, 4 May 2020, para 62, edpb.europa.eu/sites/default/files/files/file1/edpbguidelines202005consenten.pdf:

> Providing information to data subjects prior to obtaining their consent is essential in order to enable them to make informed decisions, understand what they are agreeing to, and for example exercise their right to withdraw their consent. If the controller does not provide accessible information, user control becomes illusory and consent will be an invalid basis for processing.

[274] *British Gas Trading Limited v The Data Protection Registrar* [1998] UKIT DA983492 (24 March 1998).

[275] GDPR, Art 21.

[276] GDPR, Art 22. See also the decision of the Rechtbank Amsterdam in *Unnamed applicants v Uber BV* (2021) Case C/13/692003 / HA RK 20-302, where automated decisions that the applicant drivers had used the Uber app fraudulently were reviewed and confirmed by a human employee. The court held that this human involvement meant that the decisions were not based solely on automated processing, and so there was no entitlement to an explanation under Art 22.

[277] Hongyi Wen et al, 'Exploring Recommendations under User-Controlled Data Filtering' (2018) *Proceedings of the 12th ACM Conference on Recommender Systems* 72. For a discussion of the use and limitations of data minimisation at the intersection of law and technology, see Asia J Biega and Michèle Finck, 'Reviving purpose limitation and data minimisation in data-driven systems' (20210) *Technology & Regulation* 44, 45–47.

5.3.2.2. Outcome Fairness

The second role relates to subsequent processing, which must be fair in the way in which it balances the fundamental rights and the interests of all those concerned, including third parties.[278] This is clearly an obligation to achieve outcome fairness – the law does not focus on the controller's actions, no matter how well-intentioned, but rather on the effects of the processing on the data subject. Recital 39 of the GDPR offers guidance to controllers about factors they should bear in mind when undertaking this balancing process,[279] but does not set out a definitive checklist which AI developers might use.

The CJEU has not yet been asked to examine a case under the fair processing provisions of GDPR, Article 5(1)(a) but has discussed proportionality, a similar rights balancing exercise, in the *Digital Rights Ireland* and *Schrems (I and II)* decisions. In *Digital Rights Ireland*[280] the question before the Court was whether the Data Retention Directive 2006[281] was invalidly made because its communications data retention provisions conflicted with the EU Charter of Fundamental Rights. Both the *Schrems I*[282] and *Schrems II*[283] cases challenged the validity of European Commission decisions recognising the voluntary US data protection schemes (the Safe Harbor and the Privacy Shield respectively). In all three cases the Court held that the fundamental question was whether the legislation or decision was proportional, in broad terms that it interfered with individual rights no further than was necessary to achieve a legitimate legislative purpose.

[278] Winston J Maxwell, 'Principles-based Regulation of Personal Data' (2015) 5 *International Data Privacy Law* 205, 210.

[279] The second half of recital 39 (the first relating to transparency) tells controllers that:

> In particular, the specific purposes for which personal data are processed should be explicit and legitimate and determined at the time of the collection of the personal data. The personal data should be adequate, relevant and limited to what is necessary for the purposes for which they are processed. This requires, in particular, ensuring that the period for which the personal data are stored is limited to a strict minimum. Personal data should be processed only if the purpose of the processing could not reasonably be fulfilled by other means. In order to ensure that the personal data are not kept longer than necessary, time limits should be established by the controller for erasure or for a periodic review. Every reasonable step should be taken to ensure that personal data which are inaccurate are rectified or deleted. Personal data should be processed in a manner that ensures appropriate security and confidentiality of the personal data, including for preventing unauthorised access to or use of personal data and the equipment used for the processing.

[280] *Digital Rights Ireland Ltd v Minister for Communications, Marine and Natural Resources and Others and Kärntner Landesregierung and Others*, Joined cases C-293/12 and C-594/12 (8 April 2014).

[281] Directive 2006/24/EC of the European Parliament and of the Council of 15 March 2006 on the retention of data generated or processed in connection with the provision of publicly available electronic communications services or of public communications networks and amending Directive 2002/58/EC, OJ 2006 L 105/54 (13 April 2006).

[282] *Maximillian Schrems v Data Protection Commissioner*, Case C-362-14 (6 October 2015).

[283] *Data Protection Commissioner v Facebook Ireland Limited and Maximillian Schrems*, Case C-311/18 (16 July 2020).

Consolidating the points made in the three judgments, proportionality can be explained as follows:

- The interference with rights must be necessary to achieve the legislative objectives, and should recognise that the permissible level of interference might not be the same for different types of data.
- The legislation or decision should consider safeguards for those rights and implement them. The safeguards must be likely to be effective, and should where relevant include safeguards to protect against third party actions which might damage fundamental rights. An important safeguard would be limitations on the processing of the data in question and on the time for which it would be stored or processed.
- A proportional interference will not deprive the individual of a remedy for rights violations.

These are all matters to be considered before the potentially rights interfering activity (for data protection, its processing including the transfer of data to a third country) is carried out, and are therefore best viewed as aspects of procedural fairness. They provide little assistance in determining whether the outcomes of processing are fair.

The UK ICO guidance on AI fairness[284] focuses more obviously on outcome fairness, and suggests that there are three fairness factors which AI developers need to consider, particularly when drawing inferences about individuals. In those cases the AI's processing and predictions should:

- Be sufficiently accurate for the purpose. The guidance notes that 'statistical accuracy' calculated across the AI's decisions as a whole may not equate with accuracy for each individual subject, which is what is required by data protection law.
- Avoid unlawful discrimination. The ICO's document discussing human bias and AI[285] explains these issues in more detail.
- Process personal data in accordance with the data subject's reasonable expectations.

The guidance goes on to offer suggestions about how AI developers can help avoid inaccuracy and discrimination, primarily through appropriate training and testing processes for the AU, but ultimately the guidance always returns to the undefined obligation of fairness as an outcome which must be achieved.

[284] ICO Guidance, 'What do we need to do to ensure lawfulness, fairness, and transparency in AI systems?', ico.org.uk/for-organisations/guide-to-data-protection/key-dp-themes/guidance-on-ai-and-data-protection/how-do-we-ensure-fairness-in-ai (part of the wider ICO 'Guidance on AI and data protection', ico.org.uk/for-organisations/guide-to-data-protection/key-dp-themes/guidance-on-ai-and-data-protection).

[285] ICO, 'Human bias and discrimination in AI systems' (blog post), ico.org.uk/about-the-ico/media-centre/ai-blog-human-bias-and-discrimination-in-ai-systems.

Achieving the outcome fairness which data protection law demands is clearly a major challenge for AI. There has been extensive research by the technical community into building fairness into AI (see chapter seven), and this has primarily focused on collective fairness by comparing outcomes for groups and seeking comparability in the statistical measurement of those outcomes.[286] The ICO's human bias and AI blog summarises these approaches:[287]

1. **'Anti-classification'** – according to which a model is fair if it excludes protected characteristics from consideration. Some anti-classification approaches also try to identify and exclude proxies for protected characteristics (eg attendance at a single-sex school). This can be impractical as removing all possible proxies may leave very few predictively useful features. Also, it is often hard to know whether a particular variable is a proxy for a protected characteristic without further data collection and analysis.
2. **Outcome and error parity**, which compares how members of different protected groups are treated by the model.
   - *Outcome parity*: a model is fair if it gives equal numbers of positive or negative outcomes to different groups.
   - *Error parity*: a model is fair if it gives equal numbers of errors to different groups. Error parity can be broken down into parity of false positives or false negatives (see our Accuracy[288] blog post for more details).
3. **Equal calibration** – Calibration measures how closely the model's estimation of the likelihood of something happening matches the actual frequency of the event happening. According to 'equal calibration' a model is fair if it is equally calibrated between members of different protected groups … (NB: Equal calibration does not necessarily require good calibration, only that any imperfections should affect protected groups equally).

However, the law requires the processing of personal data to be fair to the individual data subject, not merely the groups of which that subject is a member, and researchers are investigating various techniques which attempt to assess this (see section 7.5). The main problem which those researchers face is that individual fairness requires an assessment of the decision in its particular context, and because fairness at law is undefined it is impossible to identify in advance what context needs to be captured for processing by the AI.

Context, differing in every case, is what makes a definition of fairness so difficult. The few UK cases on data protection fairness illustrate this clearly.

[286] As a simplistic example, a recruitment assessment AI might be assessed by measuring the proportion of male and female applicants who are selected, assuming that a 50% selection rate for each would mean the system was fair in that respect. The actual tests for collective fairness are more sophisticated than this simple example.

[287] ICO (n 285). For a more extensive review see Sahil Verma and Julia Rubin, 'Fairness Definitions Explained' in *Fair-Ware'18*: IEEE/ACM International Workshop on Software Fairness, Gothenburg (29 May 2018), doi.org/10.1145/3194770.3194776.

[288] ico.org.uk/about-the-ico/media-centre/ai-blog-accuracy-of-ai-system-outputs-and-performance-measures.

While fairness is a risk-balancing exercise, differing contexts bring different interests into play and lead to different balancing outcomes. Early cases such as *CNN*[289] and *Infolink*[290] seemed to suggest that that the interests of the data subject ought to be given the highest consideration, but later cases have held that other interests could weigh stronger in a fairness decision.[291] However, the *Schrems* decisions[292] suggest that in a proportionality assessment the fundamental rights of an individual can outweigh even the strong economic and public interest in the exchange of data between the EU and the US. Clearly, there are no hard and fast rules to guide interest balancing.

That said, even though data protection law does not define fairness it does go some way towards identifying elements which will constitute unfairness. The most obvious of these is unlawful discrimination (see section 5.3.1 above), which is specifically mentioned as an aspect of unfairness throughout the GDPR.

Unlawful discrimination is an example of bias, favouring the members of one group over the members of another. A biased decision is not automatically unfair, because the purpose of many decision-making systems, whether human or AI, is to decide whether to favour one individual over others. A recruitment selection system will obviously be biased against less-well-qualified applicants. A system to assess patients for surgery will be biased against those who are least likely to benefit. A lender is clearly entitled to be biased against those with a poor repayment record for past loans. Where bias is definitively unfair, however, is if it amounts to unlawful discrimination, which is bias on the grounds of a protected characteristic. Most jurisdictions have some concept of unlawful discrimination, and their legislation lists those protected characteristics.[293] Even so, under UK and EU law discrimination is not unlawful, and thus potentially fair, if it is objectively justifiable. For example, a person whose physical disability includes spells of dizziness which impair their balance is clearly not safe to be employed as a roofer, and discrimination against them would be justifiable on that ground.

A second type of unfairness might be deduced from the UK case law, which seems to decide that it is unfair to make a decision on grounds which have no causal connection to the outcome.[294] If this deduction is correct it may have

[289] See n 297 and text below.
[290] See n 298 and text below.
[291] These include:

- The public interest – *House of Commons v Information Commissioner and Norman Baker MP* [2007] UKIT EA00060015 (16 January 2007); *X*, Case C-486/12 (12 December 2013).
- The controller's interests – *Johnson v The Medical Defence Union Ltd (2)* (n 295 and accompanying text); *College van burgemeester en wethouders van Rotterdam v MEE Rijkeboer*, Case C-553/07 (7 May 2009).
- Third party interests – *Promusicae v Telefonica*, Case C-275/06 (29 January 2008).

[292] nn 282 and 283.
[293] For the UK see Equality Act 2010, ss 4–12.
[294] We have already seen that UK administrative law considers such a decision to be in breach of the rules of natural justice – see section 5.2.

important implications for AI fairness, because the machine learning used to train an AI identifies correlations between the factors it uses and the 'correct' decision. As is well-known, correlation does not equal causation, which means that machine learning is at risk of basing its decision on correlations with underlying factors which have no causal link with the outcome.

The causation argument was specifically advanced in *Johnson v The Medical Defence Union Ltd (2)*.[295] In that case, the MDU referred Johnson to its risk review process, which decided that his membership should not be renewed. This required him to obtain alternative professional indemnity insurance at a higher cost. The review decision was heavily influenced by a risk-scoring system which increased his risk score because of allegations of poor practice made against him by patients. Johnson argued that this data processing was unfair because many of the allegations had either been dismissed or not proceeded with by the complainant, and that those particular allegations could therefore not have created any risk of claims and so should have been excluded when calculating his risk score. The Court rejected this argument because the MDU produced evidence that the scoring of allegations was relevant not to the risk from each allegation, but rather to identifying patterns of behaviour which might lead to claims even if those claims did not eventuate. Such a pattern of behaviour was shown to cause increased risk, and the correlation with the volume of allegations identified the relevant behaviour. It seems clear, though, that if a causal link had not been shown then the court would have decided the processing was unfair.[296]

This decision can be contrasted with the *CNN*[297] and *Infolink*[298] decisions. Both cases related to credit-scoring reports where the score was in part based on the credit histories of other individuals currently or previously living at the credit applicant's address. The defendants were able to show that there was a correlation at a group level between these third-party credit histories and the risk of an applicant defaulting. In both cases the judgment of the Data Protection Tribunal stressed that the interests of the individual data subject was the most important consideration when assessing fairness of processing, and that processing data which had no link to the applicant was therefore unfair. Although the Tribunal did not describe the lack of any link in terms of causation, the discussion of the evidence of statistical correlation in *CNN* was framed around the 'relevance' of

[295] [2006] EWHC 321 (Ch) (3 March 2006).

[296] A second finding of the court was that including third-party supplied information in the scoring was unfair because Johnson was given no chance to correct any errors. This too hints at causality, as a decision grounded in incorrect information is at risk of being wrong for that reason.

[297] *CCN Systems Ltd CCN Credit Systems Ltd v Data Protection Registrar* [1991] UKIT DA9025498 (15 February 1991).

[298] *Infolink Ltd v Data Protection Registrar* [1992] UKIT DA9025496 (28 February 1992).

using the data.[299] It seems reasonable to deduce that the Tribunal believed the credit history of an unconnected individual could not be causally linked to any increased risk on the part of the individual applicant.

Correlation is also relevant to indirect unlawful discrimination, where the decision is based on a factor or factors which are an unrecognised proxy for a protected characteristic. Examples might include religion or postcode as a proxy for race, or breaks in employment history, which could be a proxy for pregnancy. Indirect discrimination will have occurred if the underlying factor turns out to be the fundamental reason for the decision, which brings us back to causation again.

AI fairness researchers are aware of this issue and are seeking ways to test for unidentified discriminatory proxies.[300] The challenge lies in identifying the underlying factors to which the proxies correlate and their causal link to the decision, both to avoid discrimination and to prevent other kinds of unfairness. Advances in XAI (eXplainable AI) may be useful in identifying these links by elucidating which inputs to the AI were most influential in reaching the final outcome.[301]

At the end of the day, though, it seems impossible to define fairness more precisely than this: fairness means deciding the balance of interests in the same way that an averagely fair human would decide.[302]

This definitional vagueness may be less of a problem for AI than it at first seems. An AI developed through machine learning will be trained on a dataset of real-life decisions by humans. If that dataset can be cleansed of unfair decisions, the trained AI will in most cases balance the interests at stake in a very similar way to an averagely fair human, and thus meet our vague definition above.

[299] *CNN* (n 297) [33]–[37]. See also *Google Spain SL and Google In. v Agencia Española de Protección de Datos (AEPD) and Mario Costeja González*, Case C-131/12 (13 May 2014) [92]–[94], [98], holding that processing can become incompatible with the data subject's rights if it is 'inadequate, irrelevant or excessive in relation to the purposes of the processing', and time can render it so. The public interest can override the subject's rights, but in this case the time lapse of 16 years meant it was insufficient to do. Although the court does not state this directly, the inference must be that age can render data insufficiently relevant.

[300] An accessible overview can be found in Reuben Binns, 'On the apparent conflict between individual and group fairness' (2020) *Proceedings of the 2020 Conference on Fairness, Accountability, and Transparency* 514.

[301] See further Reed, Grieman and Early (n 70); Francesco Sovrano, Salvatore Sapienza, Monica Palmirani and Fabio Vitali, 'Metrics, Explainability and the European AI Act proposal (2022) *5 J* 126.

[302] See also Clifford and Ausloos (n 267) 142:

> The very point of fair balancing is that it cannot be defined at an abstract theoretical level as its application is defined by the particular context in which it is applied … Indeed, whether one fundamental right or freedom overrides another cannot be determined in absolute terms but only on a case-by-case basis … With this in mind, the only type of potential guidance on fair balancing of fundamental rights and freedoms would relate to its practical operation (eg through the definition of procedural safeguards), rather than its substance (defining lists of criteria).

Additionally, Article 22 of the GDPR requires the possibility of appeal to a human if the AI decision involves the processing of personal data, because using personal data for automated decision-making is only permissible if it falls within one of the exceptions in Article 22(2).[303] These are that the decision:

(a) is necessary for entering into, or performance of, a contract between the data subject and a data controller;
(b) is authorised by Union or Member State law to which the controller is subject and which also lays down suitable measures to safeguard the data subject's rights and freedoms and legitimate interests; or
(c) is based on the data subject's explicit consent.

In all of these cases the decision-making must incorporate: 'suitable measures to safeguard the data subject's rights and freedoms and legitimate interests, [including] at least the right to obtain human intervention on the part of the controller, to express his or her point of view and to contest the decision.'[304]

In effect, the law here requires the AI user to institute some system to appeal the AI's decision to a human decision-maker, and the rules of natural justice are highly likely to apply to such an appeal. Appeals are only likely to be made if a decision subject perceives the decision as unfair. Such an appeal can be useful in improving an AI's performance, on the assumption that if the appeal is upheld the AI will be retrained on training data which incorporates the appeal decision's findings. The consequence should then be that the unfairness element identified by the appeal ought to be eliminated from future decisions by the AI.

In practice, AI developers and users would prefer some flexibility about whether they need to incorporate appeals into their AI decision systems. For example, an appeal against an AI decision not to shortlist a job candidate would probably not be decided until after the position had been filled, and thus might seem pointless. However, it would still have the beneficial effect of identifying any unfairness in the AI's decision and enabling it to be rectified. Whether that flexibility will be allowed, though, depends on the enforcement policy of each national data protection regulator.

[303] This is because of the general prohibition in GDPR, Art 22(1):

> The data subject shall have the right not to be subject to a decision based solely on automated processing, including profiling, which produces legal effects concerning him or her or similarly significantly affects him or her.

This general prohibition might seem odd in a world where so much automated decision-making takes place, but it originated nearly 40 years ago in the EU Data Protection Directive 1995 (Directive 95/46/EC on the protection of individuals with regard to the processing of personal data and on the free movement of such data, OJ L281/31, 23 November 1995) Art 15. At that time, automated decision-making was a purely deterministic matter using simple decision trees for such matters as credit scoring. Today's AI decision-making systems are built on advances in neural networks and machine learning, which were made in the 1990s and 2000s respectively – see Alexander L Fradkov, 'Early History of Machine Learning' (2020) 53-2 *IFAC PapersOnLine* 1385.

[304] GDPR, Art 22(3).

5.3.2.3. AI Development and Data Protection Fairness

The analysis so far tells us that data protection law requires two kinds of fairness. Procedural fairness is specified in some detail in the legislation, case law and guidance, and it should be comparatively easy for AI developers to build this into their processes. Outcome fairness cannot be specified beyond saying that: (a) it requires a rights balancing exercise which is highly context-dependent; (b) unlawful discrimination is by definition unfair; and (c) that, probably, deciding on the basis of factors which are proxies for matters which can have no causal link to the decision might be unfair. This means that the best that AI developers can do to achieve outcome fairness is to establish processes which at least reduce the unfairness risk, and perhaps establish appeals to humans to identify unfairness and rectify it.

AI fairness researchers are already working on fairness processes, primarily relating to cleansing bias from training datasets and testing for bias.[305] These help to achieve collective fairness as between groups but, because they cannot capture the wider context of individual decisions, may do little to enhance individual fairness.

The ICO guidance[306] suggests that best practice for AI developers is to articulate, document and justify the choices they make during the design, training and testing of their systems, so as to demonstrate that they have made best efforts to achieve fairness at an individual level. An important element of that justification will be how the developer selected a fair and representative training set, and how it tests for unfair decisions both before deployment of the AI and in its use. Compliance with the law still remains the responsibility of the data controller, but the ICO's practice is that where best efforts have been made to comply, discussion to remedy any subsequently identified defects is preferable to imposing fines.[307]

Hacker suggests[308] that the GDPR's requirement for data protection impact assessments (DPIAs)[309] could be leveraged by data protection regulators to achieve more effective bias prevention. The analysis above suggests that this might also help in achieving other aspects of outcome fairness. DPIAs would need to use XAI tools to discover underlying hidden factors which a court might see as lacking causality. They would also need to identify as many as possible of the diverse rights and interests at stake, which might be achieved to an extent through a series of individual case studies rather than simply attempting statistical analysis of the distribution of decision outcomes. This will be challenging for AI developers

[305] For an accessible overview see Philipp Hacker, 'Teaching fairness to artificial intelligence: existing and novel strategies against algorithmic discrimination under EU law' (2018) 55 *Common Market Law Review* 1143, 1175–77.

[306] n 284.

[307] ICO, *Regulatory Action Policy* 23–25, ico.org.uk/media/about-the-ico/documents/2259467/regulatory-action-policy.pdf.

[308] Hacker (n 305) 1174–83.

[309] GDPR, Arts 35–36.

because of the importance of context – capturing a full enough context for a decision requires the collection of data which is not obviously necessary for making decisions in other cases. However, what is necessary cannot be known until the decision is questioned, and a policy of broad and deep collection of personal data appears to be prohibited by data protection law.[310]

All these measures may enhance the fairness of AI decision-making but, as we have seen, they cannot be expected to ensure that every decision is fair in terms of its outcome. This is almost certainly true for human decision-making as well, and so it may be worth trading off fairness perfection against the other benefits of adopting AI. Appeal processes can be introduced to rectify the minority of unfair decisions, as is common for human decision-making systems. It may well be that the best we can hope to achieve is for AI producers to do their best to achieve fairness, guided by recommendations from regulators and other sources of expertise, particularly in the social fairness arena.

5.3.3. Outcome Fairness in Financial Services

For nearly 50 years the UK has imposed obligations on financial services firms to treat their customers fairly, beginning with the insurance sector and expanding over time to encompass all regulated financial services firms.[311] A general obligation of fairness was introduced by the Financial Services Authority (FSA) in 2001[312] and its application has been reviewed and nuanced over the years,[313] most recently in 2021 through guidance on the special case of vulnerable customers.[314]

The obligation is set out in the Financial Conduct Authority's (FCA) Principles for Business,[315] particularly Principle 6: 'Customers' interests: A firm must pay due regard to the interests of its customers and treat them fairly.'

From 2023 there is a new duty for consumers, Principle 12: 'A *firm* must act to deliver good outcomes for *retail customers*'.[316] This is intended to subsume

[310] See eg GDPR, Art 5(1)(c). However, Asia Biega and Michèle Finck, 'Reviving purpose limitation and data minimisation in data-driven systems' (2021) *Technology and Regulation* 44, 59 suggest that this apparent paradox may not exist: 'the adequacy requirement can itself be read as requiring the collection of more data to comply with considerations such as fairness. From a legal perspective, however, what is relevant, adequate and necessary should be determined on a case-by-case basis, taking into account contextual factors.'

[311] Andromachi Georgosouli, 'The FSA's "Treating Customers Fairly" (TCF) initiative: What is so good about it and why it may not work' (2011) 38 *Journal of Law and Society* 405, 408–09.

[312] FSA, 'Treating Customers Fairly after the point of sale', FSA Discussion Paper 7 (2001).

[313] For a helpful account of its development in the early years see Nathan Willmott and Polly James, 'Treating customers fairly: meeting the FSA's expectations in a more principles-based regulatory regime' (2008) 55 *Compliance Officer Bulletin* 1.

[314] FCA, 'Guidance for firms on the fair treatment of vulnerable customers', FCA Finalised Guidance FG21/1 (2021).

[315] FCA Handbook PRIN 2.1, www.handbook.fca.org.uk/handbook/PRIN/2/1.html.

[316] Consumer Duty Instrument 2022 (FCA 2022/31), Annex B 2.1.1.

and extend the scope of Principle 6,[317] which no longer applies to retail customers, but existing guidance on fairness will still remain relevant because a good outcome includes it being fair to the customer.

Currently, Principle 6 is drafted in such a way that it imposes an obligation to achieve a defined set of fairness *outcomes*.[318] However, it does not specify what steps a firm should take to achieve these outcomes. The guidance issued by the FCA[319] indicates the factors which a firm should consider when assessing the fairness of its products, services and processes, but does not provide any assistance in the qualitative judgement about whether fairness has actually been achieved. However, the uncertainty this produces is alleviated by the policy set out in those guidelines:

> If a person acts in accordance with the Guide in the circumstances contemplated by the Guide, then the FCA will not take action against that person in relation to the aspects of the rules to which the Guide relates.[320]

A firm which followed the guidance but was judged not to have achieved compliance with Principle 6 might thus be required by the FCA to make changes, but would not face other sanctions.

Georgosouli summarises the workings of the regulatory regime concisely:

> [Firms] are induced to develop plans and monitoring systems for the delivery of certain public policy objectives. Compliance is determined by whether a firm has acceptable plans, not on the basis of detailed adherence to prescriptions or outcomes of the production process. Process-oriented regulation focuses on the firms' engagement in a process of comprehensive self-evaluation, design, and management of their business. Finally, performance-based regulation constitutes an extension of principles-based

[317] FCA, 'A new Consumer Duty', FCA Policy Statement PS22/9 (2022), 4.15.

[318] And six sub-outcomes, see www.fca.org.uk/firms/fair-treatment-customers (originally issued by the FSA 2006, last updated 24 March 2021):

> There are six consumer outcomes that firms should strive to achieve to ensure fair treatment of customers. These remain core to what we expect of firms.
>
> - **Outcome 1:** Consumers can be confident they are dealing with firms where the fair treatment of customers is central to the corporate culture.
> - **Outcome 2:** Products and services marketed and sold in the retail market are designed to meet the needs of identified consumer groups and are targeted accordingly.
> - **Outcome 3:** Consumers are provided with clear information and are kept appropriately informed before, during and after the point of sale.
> - **Outcome 4:** Where consumers receive advice, the advice is suitable and takes account of their circumstances.
> - **Outcome 5:** Consumers are provided with products that perform as firms have led them to expect, and the associated service is of an acceptable standard and as they have been led to expect.
> - **Outcome 6:** Consumers do not face unreasonable post-sale barriers imposed by firms to change product, switch provider, submit a claim or make a complaint.

[319] FCA, 'The Responsibilities of Providers and Distributors for the Fair Treatment of Customers (RPPD)', www.handbook.fca.org.uk/handbook/document/RPPDFCA20130401.pdf.

[320] ibid, para 1.3.

> regulation in the sense that it focuses on the attainment of outcomes and leaves it to the regulated population to decide how best these can be achieved.[321]

Most importantly from the perspective of AI developers, the factors which should be identified and used to label training data for machine learning are not specified – each firm needs to decide those factors for itself in the light of the guidance given by the FCA.

In theory this approach should work. The FCA mode of regulation has been described as 'conversational', a system in which regulatees (here, financial services firms) employ internal specialists who meet regularly with the regulator and each other, gradually developing a collective understanding of what compliance requires and, for each regulatee, a more individual understanding.[322] Thus a financial services firm which wished to develop AI to assist in treating customers fairly would, through internal discussion, come up with a scheme for the AI's design, its training and its testing which, in the firm's opinion, seemed to meet the regulatory objective and fit the needs of the firm and its customers. That scheme would be discussed with the FCA and modified accordingly, at which point the firm would have some assurance that it had complied with the regulatory obligation. Of course, the decisions of the AI would need to be monitored, and potentially unfair outcomes discussed with the FCA to decide what modifications to the AI were needed. In effect, the regulator would become part of the AI development team.

This is a purely theoretical description, however. In practice, regulators can never have enough resources to engage in the kinds of continuing discussions envisaged by the theory. Regulators such as the FCA are now implementing 'regulatory sandboxes' to assist firms in assessing innovative technologies.[323] If a firm applies successfully, the FCA promises to devote a dedicated case manager to engage in the kinds of detailed conversations envisaged by the theory, and eventually to approve the technology as compliant with regulation. However, there are insufficient resources to accept all applications, and so many firms are thrown back to rely on their own judgement.

Additionally, there are three known problems with this regulatory model.[324] First, it requires detailed guidance from the regulator, usually published in the form of guidelines. These guidelines become more detailed over time, and tend to transform into de facto rules.[325] Regulatees respond by turning them into a compliance checklist – a phenomenon which might be welcomed by AI developers, but may not produce an AI which meets the fairness obligation. Second,

[321] Georgosouli (n 311) 411.
[322] Julia Black, *Rules and Regulators* (Oxford, Clarendon Press, 1997); Julia Black, 'Talking About Regulation' (1998) *Public Law* 77.
[323] www.fca.org.uk/firms/innovation/regulatory-sandbox.
[324] Georgosouli (n 311) 420–24.
[325] This phenomenon is known as 'rulification' – see Frederick Schauer, 'The Tyranny of Choice and the Rulification of Standards' (2004) 14 *Journal of Contemporary Legal Issues* 803.

responsibility is gradually transferred from the regulatee because the detail of the regulator's guidelines gradually removes the discretion of management to devise and implement its own solution, even though final responsibility for compliance still remains with the firm's management. These two factors push management and development teams to focus increasingly on checklist compliance[326] and to lose sight of the real objective, which is to achieve fair outcomes for customers. Finally, implementation within the regulatee firm via computing technology introduces the programmers' own beliefs and choices about fairness.[327] As we will see in chapter seven, in the field of AI fairness these beliefs and choices tend to be derived from a limited subset of philosophical approaches to fairness, and take little account of the sociological and psychological perspectives which more closely align with the ways in which customers perceive fairness.

5.3.4. Framing Outcome Fairness Obligations

The outcome fairness obligations we have just examined are all expressed in absolute terms. The regulatee is obliged to achieve that outcome – a fair decision – and any failure is a compliance breach. In practice, though, as we saw for data protection and financial services, regulators often recognise that reasonable or best efforts may be all that can realistically be demanded.

The fundamental problem with absolute regulatory obligations to achieve a fairness outcome is that they are incompatible with AI based on machine learning (ML). In simplistic terms, an ML-based AI develops a model for decision-making by identifying patterns in its training dataset. The assumption is that using this model to make decisions for cases outside the training set is likely to produce the desired result, in this case fair decisions. The AI is tested on new cases and, where it fails to produce the desired outcome, retrained until it does so. Its final performance could even be far better than a human performing the same task, but it will always fail in a minority of edge cases, and may fail in ways that a human would not have failed.[328] Transparency about the AI's 'reasoning', often put forward as a way to assess in advance the risk of potential failures so that they

[326] A further push factor is that compliance with guidelines will generally mean that the FCA is unlikely to take enforcement action if unfair outcomes are achieved, whilst taking an independent approach risks a range of FCA action including fines – Willmott and James (n 313) 29–32.

[327] This is particularly problematic because AI fairness is such a new research field in AI that technical expertise is a rare commodity. Even if a firm's AI developers possess that expertise, it is unlikely that management will have sufficient understanding to challenge their view, and regulators may similarly lack the expertise needed to perform their regulatory role.

[328] For example, the Uber autonomous vehicle crash in 2018, where the car's driving system identified that there was some obstruction in the road but failed to slow down because it had not yet identified the nature of the obstruction – see NTSB Preliminary Report HWY18MH010, www.ntsb.gov/investigations/AccidentReports/Reports/HWY18MH010-prelim.pdf; NTSB Board meeting documents, 19 November 2018, www.ntsb.gov/news/events/Pages/2019-HWY18MH010-BMG.aspx.

can be guarded against, is seen as problematic by regulatees[329] and in any event cannot solve this problem. This is because the 'reasoning' of most ML-based AIs is not human-like reasoning in cause and effect terms – rather, the best explanations that can be given are probabilistic, explaining the factors which were likely to have led the AI to assign this case to its category ('fair') and the probability that each exerted influence on the decision.[330] ML seeks out factors which correlate with the desired outcome, and we have already noted in section 5.3.2.2 that it may not be possible to identify the causal factors to which those correlations relate.

This means that an absolute regulatory obligation to achieve fair outcomes, using AI, is an obligation which is impossible to achieve. Failures are inevitable, though this is probably also true for human decision-making, but human decisions will rarely be unfair in unexpected ways. A firm might introduce an AI which increased the overall fairness of its decisions, but still be penalised for the minority which are blatantly unfair by human standards. This would be a disincentive to adopt the technology and achieve its benefits.

Two solutions seem possible if the obligation is expressed in absolute terms:

- Dealing with the issue via an enforcement policy[331] which penalises clear failures in design, training or testing, but for outcome failures only requires suitable revision of the AI to avoid those failures in the future. To an extent, the statement in the FCA guidance already does this,[332] as does the ICO's regulatory action policy.[333]
- Requiring supplementary processes to be in place to correct outcome failures,[334] and only treating a regulatory failure to achieve a fair outcome as having occurred if the AI's decisions are not corrected appropriately.

[329] See eg Bank of England and FCA, 'Machine learning in UK financial services' (October 2019) 4.4, 19, www.fca.org.uk/publication/research/research-note-on-machine-learning-in-uk-financial-services.pdf; Bank of England and FCA, 'Machine learning in UK financial services' (11 October 2022) 21–22, www.fca.org.uk/publications/research/research-note-machine-learning-uk-financial-services-2022.

[330] Reed, Grieman and Early (n 70).

[331] Sparrow argues that enforcement policy can be the most effective tool to achieve regulatory outcomes, as opposed to mere technical compliance with the wording of regulation – Malcolm Sparrow, *The Regulatory Craft* (Washington, The Brookings Institution, 2000) 4–6, 17–25 and 238–54. See also CL Ford, 'New Governance, Compliance, and Principles-Based Securities Regulation' (2008) *45 American Business Law Journal* 1.

[332] FCA guidance (nn 319 and 320).

[333] n 307.

[334] The Bank of England and FCA surveys of 2019 and 2022 (n 329) identify four such mechanisms which are in common use in the financial services sector:

- Requiring all AI decisions to be approved by humans before they are implements. This is the 'human in the loop' approach, but is likely to be expensive and unworkable for mass-market activities such as retail loan approvals.
- Alerting humans to unusual or unexpected outcomes so that they can be reviewed and if necessary corrected.
- Backup systems which can continue to perform the function if the AI appears to be producing erroneous decisions.
- 'Guardrails' to shut continuously learning AI systems down if undesired outputs start to be produced – this aims to combat 'model drift' in those AIs.

Alternatively the regulatory obligation could be expressed as requiring reasonable efforts to achieve fairness.[335] That would focus attention on the design, training and testing of the AI, and thus be achievable. This is very much the approach taken by the EIOPA *Artificial Intelligence Governance Principles*,[336] which concentrate throughout on these factors as the route to achieving ethical and trustworthy AI in financial services.

The problem with a reasonable efforts obligation is that it might not satisfy society's need for fairness in decision-making. As individual members of society we might accept intellectually that we cannot expect decision-makers to be perfect, but a mere promise to try hard is unlikely to engender the emotional response which constitutes fairness unless the decision-maker is already highly trusted on the basis of their track record. One way to resolve this conundrum is for the decision-maker to explain how they propose to achieve fair decisions, and to justify the choices they made about their mechanisms for doing so. Both the ICO and the FCA seem already, in practice, to regulate on this basis, assessing the quality of those explanations and justifications and whether they are actually implemented in the AI. As I will argue later, framing regulatory obligations of fairness in this way is likely to be more successful in achieving the regulatory aim.

5.4. Legal Fairness Concepts and AI Development

Throughout this discussion of legal and regulatory fairness it has been reiterated that there are no 'bright line' rules for establishing whether a decision is fair. Each decision must be assessed in relation to its particular context, which inevitably makes it incomparable in some respects to other decisions. AI fairness researchers might conclude, with some despair, that the test is no more than whether the decision generates the emotion of fairness[337] in the mind of the judge or regulator, and that there is therefore little assistance to be found in legal analysis.

However, there are two reasons why despair is not appropriate. First, in a large number of AI applications no duties of fairness arise. None of an autonomous vehicle's self-driving decisions need to be fair, though they should be

[335] Arguably this is the effect of Financial Services and Markets Act 2000, s 3B(1)(b), which provides:

> [A] burden or restriction which is imposed on a person, or on the carrying on of an activity, should be proportionate to the benefits, considered in general terms, which are expected to result from the imposition of that burden or restriction.

However, Principle 6 is worded in absolute terms – fairness outcomes must actually be achieved – and this will inevitably create uncertainty in AI developers and users about whether reasonable efforts are adequate for compliance.

[336] European Insurance and Occupational Pensions Authority, 'Artificial Intelligence Governance Principles: Towards Ethical and Trustworthy Artificial Intelligence in the European Insurance Sector' (2021), www.eiopa.europa.eu/document-library/report/artificial-intelligence-governance-principles-towards-ethical-anden.

[337] See n 43 and accompanying text.

safe.[338] A properly trained medical AI which analyses kidney scans for signs of disease is unlikely to detect, or fail to detect, unfairly. Private sector employers have no duty to select job applicants for interview fairly, apart from their duty not to engage in unlawful discrimination.

Second, the law recognises that the original decision-maker had access to relevant context which a reviewer of the decision (such as a judge) does not. For this reason, fairness decisions tend mainly to be reversed or declared invalid where the original decision-maker is unable to justify why that decision was fair. This tells us that for AI decisions, the main legal focus will be on whether the decision in question can be *justified* as fair, rather than attempting to assess the overall fairness of the AI.[339] Thus an important way to address the fairness demands of law and regulation is through explaining and justifying the AI's decision-making, and this (as we will see below) may not be as challenging as designing an AI to be 'objectively' fair from the outset.

5.4.1. Common Themes of Legal Fairness

There are some common themes which we can find throughout this legal fairness discussion. Understanding them might enable AI developers and users to construct AI decision-systems which are highly likely to meet the various legal and regulatory fairness requirements.

The first theme is that legal fairness is clearly based on interest balancing. If the AI's decision affects the rights or interests of an individual, it is important to consider the individual's interests as well as the decision-maker's own interests. The principle of proportionality suggests that an individual's *existing*[340] rights and interests should only be diminished as far as is necessary to achieve the decision-maker's objective. This might apply to an AI which determined the penalty for some regulatory breach, which therefore ought to assess matters such as the resources available to the defendant and the risk that the penalty might lead to business failure, and not solely the public interest in sending a signal to other regulatees and deterring future breaches. Similarly, an AI used to assess

[338] The trolley problem is often cited in this context, hypothesising that an AI driving a motor vehicle and faced with an unavoidable collision will need to make the ethical decision whether to crash into, say, either a group of young children or a leading heart surgeon walking a cute puppy. Human drivers do not have that kind of information about potential accident victims, and I suspect are focused entirely on trying to avoid the crash. An AI driving module will not have that knowledge either, nor would any sane motor safety regulator allow this kind of decision-making to be designed into it. Like the human, the AI will be fully focused on avoiding the crash. Artificial General Intelligence, if it is ever achieved, might face ethical questions of this kind but is unlikely to drive cars.

[339] Though, at least in cases where unlawful discrimination is alleged, evidence of built-in bias to the AI may be evidence that bias occurred in the particular case. Contrarily, evidence about how the AI developer attempted to control for bias will help counter allegations of unfairness.

[340] Taking something away from an individual always needs justification, while decisions not to confer a benefit can be more capricious.

how far a current employee has achieved the targets for matters such as passing probation or securing promotion will mainly be assessing how well their performance has met the employer's needs, but should also take appropriate account of factors such as illness, pregnancy and other disadvantages which might explain and mitigate any shortcomings.

How far proportionate decisions are required when allocating *opportunities* is harder to assess, but it seems likely that if these opportunities are being allocated as part of a public function then the proportionality principle should apply. An AI scoring bids under a public procurement exercise would be correct to give weight to an applicant's track record in delivery under previous public procurement contracts, but ought to consider enough other factors so that new or previously unsuccessful bidders still have a realistic chance of winning.

Training and testing an AI for proportionality will not be easy. One possible approach might be to consider, at the design stage, the kinds of interests which might be impacted by decisions and ensure that sufficient, appropriately labelled data is included in the training set so that those data play a part in the AI's decision. Testing could be via devising theoretical individuals with those interests and applying the AI to their cases. Comparison by humans of those decision outcomes is likely to be the best way to assess proportionality fairness or, more likely, to identify obvious unfairness. An explanation of how the AI's design, testing and training aims to achieve proportionality would also be useful, as would assistance from lawmakers and regulators in identifying the interests and other factors which the AI design should encompass – both these are explained further below.

The second theme is one of procedural fairness, which requires that decisions should be free of both actual bias and the suspicion of bias. AI researchers have investigated a number of techniques for identifying and removing actual bias, though it is important to note that there is a range of ways in which bias can manifest (see section 7.4), so explaining the reasoning for deciding which technique to use will be important here. Avoiding suspicions of bias is harder, because the suspicion is in the mind of an observer and thus not under the control of the AI developer. However, a convincing justification of the fairness choices made in developing the AI will be a useful way of reducing suspicion.

Bias is a broad concept which encompasses unlawful discrimination, and it is clearly important for an AI to avoid breaches of the law. Doing so requires choosing the most appropriate bias detection and removal technique, but the most appropriate technique might differ around the world because national approaches to unlawful discrimination are not identical. The most fruitful approach here might be to build into the AI a composite of unlawful bias controls, aiming to achieve legal compliance in the countries where the AI is most likely to be deployed. The remaining issue would be how to deal with reverse discrimination, designed to favour the most disadvantaged by reference to those factors. Reverse discrimination is, as we saw, permitted under South African law in certain circumstances, but might be unlawful in other countries, so an element of customisation for national differences may be necessary. Some of the techniques for achieving equality of opportunity which will be examined in chapter seven have the potential for reverse discrimination, and these will need to be avoided where that would be unlawful.

The third theme is that of causality, and we have seen court decisions which decide that the evidence on which a fair decision is made must both be accurate, and also have some logical causal connection to the decision outcome. The problem here is that machine learning identifies correlation rather than causation, and thus might appear to be making decisions on illogical grounds. An example could be ranking creditworthiness on the basis of an applicant's pattern of topping up the credit on their mobile phone (recharge). This data has been demonstrated to contain patterns which correlate well with creditworthiness and which can be used to improve lending decisions.[341] But because there is no logical causal connection between top-up patterns and creditworthiness, making a decision on those grounds is at first sight unfair. As a second example, social media credit-scoring is already in use, with commercial service providers integrating data from an applicant's online activity for use in loan decision-making,[342] and this suffers from the same apparent lack of logical connection.

AI developers will therefore need to justify the use of purely correlative data by using XAI and other analytical tools to identify the underlying factors with which the data correlate and checking their logical causal link with decisions. Alternatively, a social fairness justification for using that data might be constructed – in the case of the mobile phone top-up data, prediction of default using top-up data was more accurate than prediction using standard credit-scoring, but reduced the volume of credit granted by 80 per cent[343] Using this data might therefore be unfair to customers who would have been granted a loan under standard credit scoring, because of the causality defect. However, using the data to extend credit to customers who would otherwise have been refused could well be justifiable as fair, on the basis that it enhances social inclusion of those who are unbanked or otherwise unconnected with the sources of standard credit-scoring data.

Finally, fairness sometimes requires an explanation for the reasons behind a decision. Even where the law does not demand an explanation, producing one can have its uses – for example, recording why a candidate for a job was not chosen at interview may be critical to defending a later allegation that unlawful discrimination had taken place. From the perspective of both AI developers and users, explanations clearly play an important part in showing that decisions are likely to be proportionate, procedurally fair and unbiased, and in showing the causal link between a decision and its grounds.

[341] Alain Shema, 'Effective credit scoring using limited mobile phone data' (2019) *Proceedings of the Tenth International Conference on Information and Communication Technologies and Development*, 1.
[342] See eg seon.io/resources/social-media-credit-scoring.
[343] Shema (n 341) 9–10.

Explanations of this kind will be useful where law or regulation contain transparency provisions, such as the requirement under data protection law to inform data subjects about the kinds of processing of their personal data which will take place.[344] They will also help to convince decision subjects and wider society about the legitimacy of the AI's decision processes – as we saw in chapter three, the social acceptance of any decision-making system as fair is important in conferring legitimacy on it, and we noted in section 3.1.3 that publishing an explanation of the AI developer's fairness choices and justifying why those are likely to lead to fair decisions would go a long way towards establishing that social acceptance.

Explanations will be particularly important for AI-as-a-Service (AIaaS), where a generic AI decision-making model is made available online and the AI user adopts it (possibly with some limited reconfiguration) to automate some aspect of the user's decision-making. Any hidden biases in the AIaaS model will feed through to the AI user. Additionally, the AIaaS model will inevitably fail to take account of the context in which it is now being used, which is a critical element for making fair decisions.[345] Worse, an AI model which produces fair decisions in the context for which it was designed and trained may produce unfair decisions if used in a different context.[346] An explanation by the AIaaS provider of the fairness decisions and processes designed into the model will assist the AI user to decide whether using that model is appropriate, and also to construct their own explanation of how using the AI is justified in the user's context.

Deciding when a duty to give explanations might arise is difficult, but because fairness is based on comparison between outcomes a duty is unlikely to arise where the effect of a decision on individuals is not based on their particular characteristics. Thus I have already suggested that the driving decisions of an autonomous vehicle do not raise fairness issues, and this is because (say) a pedestrian injured by a crash suffers not because they were singled out from other potential victims because of their age or education, but only because they were in the wrong place at the wrong time. Fairness explanations will thus only be needed where the effects on individuals differ on the basis of those individuals' different characteristics, and are probably only necessary if those effects impact the individual's interests or deprive them of an opportunity.

This kind of explanation focuses on the overall fairness of the AI decision system, but the duty to explain also requires some explanation of how the individual

[344] See section 5.3.2. The ICO has produced detailed guidance on explaining AI to achieve compliance with data protection law – ico.org.uk/for-organisations/guide-to-data-protection/key-dp-themes/explaining-decisions-made-with-artificial-intelligence.

[345] Kornel Lewicki, Michelle Seng Ah Lee, Jennifer Cobbe and Jatinder Singh, 'Out of Context: Investigating the Bias and Fairness Concerns of "Artificial Intelligence as a Service"' (2023) *CHI '23: ACM Human Factors in Computing*.

[346] Andrew D Selbst, Danah Boyd, Sorelle A Friedler, Suresh Venkatasubramanian and Janet Vertesi, 'Fairness and abstraction in sociotechnical systems' (2019) *Proceedings of the Conference on Fairness, Accountability, and Transparency* 59.

outcome was decided on,[347] and in particular the grounds for the decision which, as we have seen, must be logically connected to the outcome. Producing such an explanation is challenging, even using XAI tools to identify the factors which are likely to have led the AI to make its decision,[348] and perhaps even more challenging is how to make the explanation meaningful to that majority of decision subjects which is not familiar with how AI technology works. Sossin makes an interesting argument that a conversational discussion between the decision subject and decision-maker can be the most effective way of eliciting the underlying reasons for a decision,[349] which would be both more meaningful for the subject and also help establish the legitimacy of the decision. One possibility, yet to be explored, is whether large language models like ChatGPT might be usable for this purpose, thus avoiding the substantial human input currently required to generate an individual decision explanation.

Providing explanations, particularly about the AI developer's approach to fairness in design, can help meet two further elements of legal fairness. First, it can help set the decision subject's legitimate expectations, and thus (if the explanation is accurate) achieve compliance with the obligation to meet them. Second, if there is the possibility to appeal an AI decision to a human, both kinds of explanation will help the subject to determine if an appeal has any realistic prospect of success. Additionally, the very process of crafting an explanation may also tell the developer or AI user whether some appeal process should exist at all by identifying the points at which the fairness of the AI might be open to question.

5.4.2. Lessons for Law Reform

AI developers and users might justifiably complain that the fairness obligations of law and regulation are too broad and ill-defined to help them achieve compliance, and lawmakers might equally justifiably reply that fairness is too complex a concept to define as a set of rules. However, the discussion above suggests that there are three things which might be achieved in making law and regulation.

[347] See *Bateman* (n 245) [117], holding that the reasons had to explain why the decision was arrived at, and the factors which justified that decision.

[348] See further Reed, Grieman and Early (n 70).

[349] Lorne Sossin, 'An Intimate Approach to Fairness, Impartiality and Reasonableness in Administrative Law' (2002) 27 *Queen's Law Journal* 809, 837–38 describing this conversational interaction as 'intimacy':

> If it were possible for decision-makers to discuss their internal reasons on a 'without prejudice' basis (in other words, in a fashion that could not be used to prejudice either party in a future legal hearing), insights into the reasoning of bureaucrats might become commonplace … The central flaw of the duty to provide reasons, as it now is articulated, is that it provides no standard on which to assess and evaluate reasons. As long as some indication is provided which discloses to the affected party the basis of the result in the circumstances of the case, the legal standard is met. On a standard of intimacy, by contrast, fuller, clearer, more comprehensive, genuine reasons will be preferable to the alternatives.

The first is that, at present, the case law shows that the context of a decision plays a major role in deciding what a fair decision might demand in that case. Obviously context is so variable that it cannot be specified in detail in advance, but lawmakers could identify the most salient contextual factors which should be considered in a particular type of decision, thus enabling AI developers to collect and label training data which allows the AI to consider those factors. This is already done to an extent in equality law, through its list of grounds on which it is unlawful to discriminate, and the discussion of financial services regulation in section 5.3.3 shows that regulatory guidance provides some assistance in identifying important contextual features for decisions in that sector.

However, while ensuring that appropriate input factors are reflected in the model is likely to improve fairness, it is important to remember that it is the reaction of subjects and society to outcomes which ultimately determines whether fairness has been achieved:

> Legal scholars have argued that traditional approach of scrutinising the inputs to a model is no longer effective due to the rising model complexity … For a more comprehensive analysis of whether a model meets the stakeholders' ethical criteria, it is important to look beyond the inputs and the designer's intent and assess the long-term and holistic outcome.[350]

More detail about context in sectoral regulation in particular, most likely via guidance rather than substantive law, because of its flexibility, could also help to reduce some of the contradictions in law which AI developers face. A particular problem is in data protection law, which limits the collection of personal data to that which is necessary for the disclosed purposes of processing. The contradiction here is that fairness requires decision-making data controllers to consider the broader context, and an AI can only do that if it has collected a wider range of data than that which can be predicted in advance to be *necessary* to make the decision. Potentially desirable data is, at the time of collection, not yet identifiable as necessary. If sectoral law and regulation identified relevant context elements which should be considered in making decisions, that would make it necessary for developers and users to collect that data and thus legitimate their doing so under data protection law. To some extent the explanations of fairness techniques recommended in the previous section might be used to identify purposes in detail and justify data collection to meet those purposes, but normative legal and regulatory statements about what kinds of data ought to be considered will help remove doubts about whether it is possible for a data controller to justify collecting additional data merely by listing the kinds of data which the controller *might* in future find useful to assess fairness.

Second, both AI developers and AI users will need to consider how far it would be beneficial to attempt to follow the rules of natural justice. These suggest that decisions should be unbiased and made on accurate and complete data; that the

[350] Lee et al (n 31) 541.

factors which lead to a decision should have a logical connection to the outcome; that explanations of the attempts made to achieve fairness and in some cases of the outcome itself can be useful in achieving acceptance that the AI decided fairly; and that where the decision impacts on an individual's rights some form of appeal mechanism might be needed for decisions which are perceived as unfair. All these are achievable, but need to be designed in to the AI.

Third, AI developers who are attempting to eliminate bias have a wide choice of techniques for doing so, each of which produces somewhat different outcomes. Currently they can do no more than guess which of those techniques is most likely to align with any fairness obligations they are subject to. Regulators could assist here through sectoral guidance, in particular by identifying approaches which are unlikely to achieve what the regulator is looking for.

Finally, we have seen that law and regulation which requires reasonable or best efforts to achieve fairness outcomes can be complied with by AI developers and users, but that the inherent uncertainty derived from training an AI via machine learning means that absolute fairness outcomes are not certainly achievable by AI in every case. This is because, ultimately, the test is the fairness emotion engendered in the mind of a human reviewing that particular decision, and the reviewer's emotional response cannot be predicted reliably.

An alternative approach to reasonable or best efforts is to impose an absolute obligation to achieve a fairness outcome but mitigate it through an enforcement policy which takes account of reasonable or best efforts. This can achieve the same ends although there is a real risk that it will be less effective in achieving the desired compliance. The risk arises because enforcement policies are less visible to developers and users than express efforts obligations – as an example, the obligation under UK financial services regulation to achieve fair outcomes is set out front and centre in the regulation,[351] while the enforcement policies of the FCA[352] are hidden in their guidance documents and need to be sought out by those in the know.

All these focus on the actions of the humans who develop and use AI, not on the workings of the AI itself. Even if a detailed set of codable rules for fairness could be devised, which the preceding chapters have I hope shown seems highly unlikely, there are good reasons why AI fairness regulation should not follow that path. A practical objection is that society's concept of fairness is constantly evolving, and so these detailed rules would need constant change to follow that.

The main principled objection is that this approach removes fairness from the judgement of wider society, and substitutes a regulatory proxy which will inevitably fail to match society's need for fairness. Georgosouli, discussing proposals to translate UK financial regulation into a form which enables

[351] n 318.
[352] n 320.

automated compliance, explains that the application of human judgement is fundamental to achieving regulatory aims. Those who are regulated ought not to be:

> [P]assive recipients of regulatory commands. They are expected to exercise discretion as they engage in a process of self-reflection on how they ought to run their business. Regulatory interventions are on the menu to communicate findings and interrogate business culture, and require a particular course of action. However, these interventions are meant to treat regulatees not as mere objects to be controlled, but as subjects capable of ruling themselves and of being accountable for their actions.[353]

Fairness, more than any other regulatory aim, requires accountability to wider society, because it is that wider society which ultimately decides the current meaning of fairness.

[353] Andromachi Georgosouli, 'Metarules, Judgment and the Algorithmic Future of Financial Regulation in the UK' (2023) 43 *Oxford Journal of Legal Studies* 777, 796.

6

The Human Conception of Fairness

6.1. Constituents of Fairness

At this point it is useful to bring together the different conceptions of fairness we have examined in chapters three to five, in order to understand both what AI developers and users should be aiming at and how far we can expect them to achieve it, and also the target which any AI fairness regulation should be aiming at. Fairness is complex, nuanced and highly context-dependent, so it is unrealistic to expect perfect fairness in AI decisions. The same is true of humans, although their imperfections are likely to be somewhat different in nature. Regulation cannot realistically do more than aim at achieving an acceptable level of AI fairness.

At its core, fairness is an emotional response to a decision (see chapter three). If that response is positive, the decision was fair. If negative, it was unfair. Emotional responses vary between individuals, as does their intensity, and so we can only assess the fairness of a decision by reference to the consensus response of the community or society within which that decision is made. Law and regulation are familiar with this problem, and regularly make reference to the judgement of the reasonable observer as the touchstone for whether an open-textured standard like fairness has been achieved.

The emotional response of fairness is only elicited once the decision has been made, which presents two problems for AI. It might be possible, with further research, to use a second AI to identify how far an individual decision subject might perceive the decision as fair,[354] but data from which to determine the wider consensus emotional response is unlikely to be available. More importantly, what society is looking for is sufficient assurance about the future, ie that the AI is likely to make decisions which will be perceived as fair. Thus we need to unpack the emotional response to identify its likely causal factors, which might then be usable by AI developers to achieve that assurance.

[354] Training an AI for emotion detection is to an extent achievable – see Catherine Marechal, Dariusz Mikolajewski, Krzysztof Tyburek, Piotr Prokopowicz, Lamine Bougueroua, Corinne Ancourt and Katarzyna Wegrzyn-Wolska, 'Survey on AI-Based Multimodal Methods for Emotion Detection' in Joanna Kołodziej and Horacio González-Vélez (eds), *High-performance modelling and simulation for big data applications* (Springer Online, 2019) 307 – though whether it can be extended beyond fundamental emotions such as happiness or sadness to more complex emotional responses is uncertain.

Fairness emotion seems to be a response to two main causal factors:

- Unjustified unequal treatment of like cases is generally perceived as being unfair. This is helpful, as we will see in chapter seven. However, the justifications for unequal treatment which each social group accepts here are culturally determined – for example, as we saw in section 3.1.1, in some societies power and status justify preferential treatment – and so achieving acceptably equal treatment will demand tailored approaches to meet each society's expectations. Some degree of unequal treatment to advance wider social equality seems to be generally acceptable, but again the level of acceptance varies between societies (see section 3.1.2) and can change – for example, the US Supreme Court has recently held that most forms of affirmative action in university admissions are no longer lawful.[355]
- Failure to consider fully the rights and interests of the decision subject and balance them against the wider interests of society also engenders emotions of unfairness. Law and regulation already recognise this factor in the concept of proportionality in decision-making (sections 5.2 and 5.3.2.2), and many elements of procedural fairness are aimed at achieving this kind of interest balancing.

Justifications for unequal treatment require deeper analysis because this is clearly one element of fairness which might be incorporated into AI design and training – indeed, AI developers and researchers have already developed a range of fairness techniques which focus precisely on such justifications.

6.2. Justifying Unequal Treatment

In broad terms, there are two main routes to justify unequal treatment of decision subjects. The first is that the favoured subject exhibits some merit and thus deserves more favourable treatment than other decision subjects, for example better qualifications for a job. The second is that the decision subject suffers from some disadvantage which ought to be compensated for in making the decision – in employment, this might be that the disadvantage has prevented the subject from acquiring the same qualifications as other candidates.

For AI decisions, merit is normally determined by the AI user. Criteria are set for a successful outcome, and then decision subjects assessed against those criteria. In the examples we have returned to in the discussion, these might be qualifications and experience for employment selection, likelihood to respond to treatment for our medical allocation AI, and likelihood to repay for loan granting. There is, to an extent, also an objective component to merit, in that selecting criteria

[355] *Students for Fair Admissions Inc v President and Fellows of Harvard College* (2023) 600 US 181.

which have no causal connection to the AI's aim is likely to be perceived as unfair. Thus employment selection on the basis of hair colour would usually be seen as unfair,[356] though probably not unlawful.

Disadvantage is more complex. It appears to start from a normative understanding of the benefits which all members society deserve – for example, that all children should receive a good education, or that all should have good health – and then considers those who did not receive that benefit to be disadvantaged. However, the behaviour of the individual also plays a part. Those who choose not to continue with their education in favour of work or travelling are not considered to be disadvantaged, or at least not considered worthy of compensatory treatment, and the same might apply to those whose ill-health is a result of their own voluntary behaviours. These issues were explored in chapters three and four and are discussed further below.

As we saw in section 3.1.1, merit and disadvantage are traded off against each other. This suggests that the human decision-maker is, in effect, assessing the decision subject on various merit continua and, if disadvantage reduced the subject's ability to achieve that merit, upgrading their assessment to compensate. This is closely analogous to the kind of weighted scoring against decision criteria which AI developers use, as we will see in chapter seven. If the merit which justifies a particular outcome is the starting point, that tells us that the decision needs only to take into account any disadvantage which reduced the chance to achieve that merit and for which the decision subject needs to be compensated. For example, a job shortlisting decision based on qualifications should consider upgrading an applicant whose illness prevented them obtaining a particular qualification, but would not upgrade if the illness had no effect on qualifications but produced other disadvantage which would not be relevant to the employment, such as decreased mobility for an office job.

However, this does not mean that a fair decision always compensates for disadvantage. There are two reasons why it can be fair not to do make such an adjustment.

The first is where the disadvantage prevents the decision subject from reaching a set threshold for a positive outcome. Some thresholds are objective – an individual who has been prevented by disadvantage from qualifying as a lawyer, doctor or other professional is still unqualified for a post in that profession and it is therefore not unfair to reject them for such a post. More difficult is the case of a patient whose medical condition makes it unlikely but not impossible that they will benefit from a particular treatment. The decision-maker here might want to set a threshold of likely benefit below which patients will be rejected for treatment, but that threshold setting is a matter of judgement rather than an objective fact.

[356] I say 'usually' because, for example, those choosing an actor for a biopic might consider hair of a similar colour to the person portrayed to be a relevant merit criterion, though of course wigs and/or hair dye are an alternative.

The focus here is likely to be on the fairness of the *process* by which the threshold was set – was it on the basis of consensus in the medical profession rather than an individual choice by the decision-maker, were patient representatives and others consulted, and so on.

The second reason is more complicated. Failure to compensate for disadvantage is not unfair unless the disadvantage *ought* to be compensated for, and the 'ought' here reflects that particular society's social norms. We saw earlier Tyler's finding that the norm is based on some conception of objective justice rather than mere sympathy for or empathy with the disadvantaged person.[357] This seems to derive from existential guilt[358] about the difference in advantage, leading to a norm that this injustice should be remedied by some distribution in favour of the disadvantaged person.[359] If society feels guilty about not taking remedial action, fairness requires the disadvantage to be remedied.

The origin of the norm in guilt might explain the sociological, psychological and philosophical consensus that there is a normative difference between disadvantage suffered unwillingly and disadvantage caused by that individual's voluntary choices. Society can choose to help alcoholics, even if some element of voluntary behaviour might have been involved, because we recognise that addiction overcomes free will. We generally decline to help those who just like being drunk if they can give up drinking whenever they feel like it.

Guilt might also help us understand how context affects the application of the norm. The greater the suffering of those who are disadvantaged, the more guilty the advantaged feel about not helping them. If we look back at the example in chapter two of a hunter-gatherer society distributing food, in times of plenty everyone has enough to eat and so do not suffer (other than envy) if some individuals receive a larger share. In times of famine the weak and sick are in greatest need, and so failing to feed them properly should engender guilt in the others. This might explain the research finding of Ubel et al[360] that a hypothetical medical screening test given to 100 per cent of the US Medicaid population which saved 1,000 lives was perceived as fairer than a more effective test given to only 60 per cent, but saving 1,100 lives. The guilt here is about the 40 per cent who are deprived of a chance of benefiting. The researchers' counter-example of the same tests, saving the same number of lives each, but given to 60 per cent and 40 per cent of the population respectively produced a different result – here the fairer test was the one which saved more

[357] See n 40 and accompanying text.

[358] The nature of such guilt is hard to analyse because the emotional reasoning is circular. It is unfair not to compensate for disadvantage if I would feel guilty about not doing so, and I would feel guilty because it is unfair not to do so. Perhaps one way of understanding this might be the collective view of the advantaged about how they would expect to be treated by others – would they expect mere sympathy and understanding, or would they expect others to take action to help them? Guilt would then arise from the failure to meet those expectations.

[359] Tyler (n 30), citing Leo Montada and Angela Schneider, 'Justice and emotional reactions to the disadvantaged' (1989) 3 *Social Justice Research* 313.

[360] n 19.

lives. In this second example, it is inevitable that some proportion of the group will have no chance of benefiting, which removes the guilt about not treating all equally and moves the focus to the overall benefit to those who are tested.

Before we leave the issue of disadvantage it is worth noting how it interacts with unfair discrimination law. The law sets out social groupings[361] which are likely to suffer disadvantage, but does not establish a norm that it is unfair not to compensate them for it. Rather, the law forbids imposing any further disadvantage on them by treating them less favourably than others because of their social group membership. In spite of this conceptual difference, the social groupings established by unfair discrimination law have been seized upon by AI developers as useful lists of disadvantaging characteristics to be incorporated in AI development and training – see section 7.4.

Finally, we should note the agreement among social science disciplines that fair decision-making *systems* (as opposed to individual decision-making) ought to have built into them a wider objective of making society less unequal. How far to do this is clearly a political decision which might more appropriately be imposed by law and regulation on some categories of AI producer, rather than being left to individual developers and users to decide.

6.3. Assessing Fairness

The basis of fairness is that like cases are treated alike, with apparent differences justified on the basis of rewarding merit and compensating for disadvantage. It is therefore obvious that the primary way in which a human will assess the fairness of a particular decision is by comparing its outcome with that of other decisions. The overall fairness of a decision system will be assessed in the same way, looking for consistency of outcome and reviewing the justification for decisions which appear to be outliers.

This is also the approach which AI developers have adopted, aiming at developing metrics which provide some objective measure of fairness for each decision and for the AI as a whole. However, this aim is difficult to achieve because of the complexity of the human conception of fairness.

First, deciding the basis for comparison is never objective. It is partly driven by the purpose of the AI user, and each user may have different merit factors to be used in making the decision. Employment gives obvious examples – some jobs require consistent following of set procedures, others demand initiative and creativity, some require physical strength and others high intellectual capacity, and so on. An AI selecting candidates for a tree surgeon job would use very different criteria from those for a heart surgeon.

[361] Sex, race and disability seem universal groupings in unfair discrimination law – for other groupings in different national laws, see section 5.3.1.

Using machine learning to analyse previous human decisions is one way of extracting merit factors. However, this runs the risks both of building in the irrational or emotional considerations which human decision-makers have, perhaps unconsciously, been influenced by, and also of identifying as merit factors some proxies for other characteristics like race or sex which it is unfair or unlawful to use.

Second, deciding whether to compensate for disadvantage is equally subjective. Disadvantage such as ill-health or childhood poverty, or societally determined disadvantage factors like sexual orientation, has complex consequences and thus the potential to relate to many or all of the merit factors used by the AI. Choosing which disadvantage factors should be considered by the AI, and the degree of compensation to be applied, are political decisions which reflect the normative stance of the AI developer and user.

Even if a suitable basis of comparison can be found, that does not by itself achieve an external perception of fairness. For decision subjects and the wider society which is judging the AI's fairness, the ways that the AI's decisions use merit and disadvantage are completely invisible in the outcome itself. Without some degree of explanation of these matters comparison of decisions is pure guesswork, and is at best likely to lead to suspicion that some of the AI's decisions might be unfair.

Even where satisfactory explanations are made available, comparison of outcomes by anyone other than those who operate the AI is likely to be impractical. Fairness questions tend to arise where the decision is important to the decision subject, impacting her rights and interests, and there is normally an expectation of a high degree of confidentiality for such decisions. Employers do not publish lists of rejected candidates, let alone with reasons; patients expect that details of their medical treatment will only be shared with other medical professionals; and banks do not publicise successful and failed loan applications.

Thus although in theory the fairness of an AI can be assessed by comparing its decision outcomes, in practice this is an unhelpful test for deciding as an outsider whether the AI acts fairly. Regulators, if they exist for the type of activity in which the AI operates, might make such comparisons to ensure compliance with any fairness regulation, but this will require greater technical expertise and more resources than are likely to be available to the regulator. This is true to a lesser extent for human decision-making, and explains why (as we saw in chapter three) that external assessments of fairness tend to be more strongly influenced by two visible elements – the procedures for making the decision, and any explanations which justify that decision.

Both procedure and explanations are *proxies* for fairness. If comparison of actual outcomes is not possible, fair procedures and explanations which convincingly justify why outcomes should be fair are all that is available for external judgement of the AI.

Procedural fairness as a legal concept was examined at length in section 5.1, and other aspects of procedure which influence fairness judgements were analysed in later sections of that chapter. In summary there are four main elements here:

- The decision must be made by taking account of all the relevant facts. In human decision processes which hold hearings this usually requires that the decision subject should be informed of the facts which the decision-maker is relying on, and given a chance to introduce additional facts and any relevant arguments about the appropriate outcome. This concept of a fair hearing applies by analogy where there is no actual hearing, obliging the decision-maker to search out relevant information so that the decision is properly informed.
- If any reasons for the decision are given, they must justify the outcome. In particular, basing a decision on matters which have no logical connection with the outcome will be unfair. This presents challenges to machine learning-based AI, as the decision factors analysed via the machine learning process might be proxies for underlying facts which have no logical connection to the outcome if they reflect prejudices or biases in the training set of human decisions.
- Legally fair procedure requires that the decision-maker exhibit no appearance of bias or prejudgement of the outcome. National unlawful discrimination laws set out some factors against which bias or prejudgement can be assessed, but AI developers need to be aware that these are not the only sources of unfair bias – an example might be selecting employees on the basis of their social background where that is irrelevant to their performance in the job. From the wider social perspective, what needs to be aimed at is at least the appearance that the AI can be trusted not to make biased or prejudged decisions.
- From a non-legal perspective, a procedure which allows the decision subject to interact with the decision-maker in a way which convinces them that they have been treated with dignity and respect makes the strongest contribution to that subject's acceptance of the decision as fair (see section 3.2.1). Interaction of this kind is not currently a feature of AI decision systems, but recent advances in generative AI offer the possibility that, in appropriate cases, the AI might engage in a 'conversation' with the subject, both eliciting the relevant facts on which to base the decision and explaining and justifying the outcome. The greater the impact of the AI's decision on the rights and interests of the subject, the more useful such an interaction would be in achieving a perception of fairness.

Finally, explanations and justifications of the ways in which the AI decision system attempts to achieve a fair outcome will play a highly significant role in external fairness judgements. The purpose of these explanations is to engender the emotion of fairness in decision subjects and external observers, and so they need to be devised appropriately for different audiences. Regulators will be interested in detailed explanations, and have the capacity to analyse and evaluate them, while private citizens are more likely to respond to short and simple explanations.

Decisions which elicit no emotional fairness response, such as the driving decisions of an autonomous vehicle, require no fairness explanation,[362] while decisions which have a substantial impact on their subjects clearly require more explanation and strong justification.[363] An explanation about the design choices for the AI, identifying the fairness techniques adopted and justifying how they are likely to help in achieving fair decisions, will provide assurance to external observers about the overall fairness of the AI decision system. Providing explanations and justification for individual decisions is more complex and expensive, and so the AI developer and user will need to decide whether doing so will be useful in achieving fairness judgements – the more serious the consequences of the decision, the more likely this will be needed.

6.3.1. Fairness Control in AI Development and Use

There are numerous reasons for AI developers and users to want their AI decisions to be substantively fair, and also to be accepted as fair. Unfair decisions are often economically irrational as well – for example, refusing to consider certain applicants as potential employees because of some irrelevant bias reduces the pool of talent from which employees can be drawn. If law and regulation begin to impose fairness obligations on AI decision systems, non-compliance will lead to sanctions. Perhaps most importantly, the wider social acceptance of a decision system as fair helps to establish its legitimacy and thus reduces opposition to its use.

However, those reasons on their own may not be enough to achieve fair AI decision-making. Law and regulation is an obvious source of incentive, and at the time of writing the pressure to make such laws is increasing because of concerns raised by generative AI. Fairness is an obvious regulatory concern.

I have, though, argued elsewhere that it is wrong, both practically and in principle, to impose regulation on technology producers and users with which they cannot comply.[364] The question which we need to answer, therefore, is what we as human outsiders can reasonably expect developers and users to do in an attempt to achieve a close match between the working of their AI and the human conception of fairness.

It is helpful here to analyse which elements of the human conception of fairness identified in this chapter six might be addressed at the AI development (design,

[362] These decisions will, of course, require explanations about other matters such as safety.

[363] When suspicions were raised that the COMPAS system, designed to assist judges in making probation decisions, might have been used to make sentencing decisions, the Supreme Court of Wisconsin issued an extensive opinion on the safeguards which would be necessary in such cases. The most important of these was that decisions about the liberty of individuals should not be made solely on the basis of a machine learning technology which cannot give a satisfactory account of the process by which its recommendations were produced – *State of Wisconsin v Loomis* (2016) WI 68, [87]–[92].

[364] Chris Reed, 'How to Make Bad Law: Lessons from Cyberspace' (2010) 73 *Modern Law Review* 903; Chris Reed, *Making Laws for Cyberspace* (Oxford, Oxford University Press, 2012).

training and testing) stages, and which have to be addressed at the time of, and subsequent to, implementation of the AI to make decisions.

Most of the actions which humans might expect AI developers and users to take will need to be built into the AI system during its development. The most important action, ultimately undertaken at implementation, is to explain and justify the fairness choices made during development. This means that at all stages those choices must be identified and articulated, and thought given to how they will be justified. Reverse analysis of an AI built without doing this is likely to produce a highly unsatisfactory explanation.

That development explanation will need to cover four main areas:

- The aim of the AI's decision-making, with a justification for the merit criteria which are adopted for that purpose.
- How the AI aims to achieve the core fairness characteristic of equal treatment with appropriate allowance for disadvantage.
- How far the rights and interests of decision subjects have been identified and balanced appropriately against the interests of the AI user and others.
- The fairness of the procedures for decision-making which have been designed and trained into the AI.

Ideally the development explanation would also clarify how the AI's development enables it to modify its decisions to take account of changes in the context. But as we have seen, the effect of context on fairness is highly determined by a society's culture, and so it may not be appropriate for society to demand too much here unless there is some collective articulation of how it should be dealt with through law and regulation.

As we will see in chapter seven, equal treatment requires the developer to decide on appropriate training data labels for matters such as merit, disadvantage and unlawful discrimination factors, and ways in which unequal decision outcomes can be identified and corrected by retraining the AI. Without external guidance such as that which law and regulation might provide, the best that society can demand is that these choices have some identified basis in research, theory or empirical analysis, and that the choices can be justified in terms of achieving fairness. Society can also expect the AI development to take account of the risks of using machine learning as the source for these matters, such as embedding existing human biases and identifying proxies rather than genuinely causal factors, and to take steps to control those risks.

Where the AI's decision will potentially have a serious impact on the rights and interests of decision subjects, AI producers can be expected to identify those rights and interests and seek to balance them appropriately against other rights and interests. This will not always be necessary for fairness – for example, in an AI credit granting system the applicant has no right to a loan, and if refused can always apply to a different lender. However, in our medical treatment allocation AI this question of rights and interests definitely plays a part. The patient may have no legal right to treatment, but does have a legitimate expectation of receiving good treatment, based on matters such as the existence of a social health system, the medical ethics obligations of health professionals, and so on. The consequences for the patient will clearly be serious if treatment is refused. The difficulty here lies in deciding the appropriate form of rights and interest balancing. Without external guidance, the best which society can expect is that the AI developer will make those choices itself, and is able to produce a convincing justification for them. One possible technique for improving and justifying this balancing exercise might be some element of human review of decision outcomes as part of the AI's testing processes.

Most of the elements of procedural fairness also need to be incorporated at the AI development stage. Obtaining full and accurate data for the AI to work on is a known problem in the field, with a range of solutions. However, additional thought might need to be given to systems for obtaining relevant supplementary data from humans, which might include the decision subject and others with relevant knowledge, such as medics in the treatment allocation example.

Ensuring there is a logical connection between the factors which influence the AI's decision and the outcome is more challenging. For AIs based on machine learning, this will require the developer to use XAI tools to identify those factors and their contributions. Once known, human analysis will probably be necessary to decide whether they are appropriately causal and to construct the explanation and justification for their use by the AI. Factors which are apparently non-causal might well be proxies for underlying factors which *are* causal, and could therefore be justified if the relationship can be discovered.

Ensuring the AI is unbiased and does not prejudge the outcome is unlikely to be a real issue, because the equal treatment decisions already discussed should prevent it arising. However, the aim of the decision-making will inevitably have some element of bias or prejudgement built in to it, because the whole purpose of the AI is to decide outcomes to fit that aim. Explanation of the aim, and how it shapes the decision-making, will therefore form an important part of the overall development explanation.

There are two elements of procedural fairness which society can probably *not* expect to be incorporated into AI decision systems:

- Whether individual outcomes require explanation and justification in order to convince the subject and others that the decision was fair. AI developers and user might choose to attempt this, but extracting an explanation for each individual decision is costly and might not produce a satisfactory answer.[365]

[365] For example, Vale et al argue that XAI techniques are only suitable for identifying some types of bias in AI decisions, and that historic or social bias derived from training on prior human decisions may remain hidden: Daniel Vale, Ali El-Sharif and Muhammed Ali, 'Explainable artificial intelligence (XAI) post-hoc explainability methods: Risks and limitations in non-discrimination law' (2022) 2 *AI and Ethics* 815.

How to generate this kind of explanation is an ongoing field of research. Over time law and regulation might demand such explanations in particular AI applications, and if so will ideally give guidance as to what will be acceptable in those cases.

- Giving the decision subject the opportunity to interact with the AI would have beneficial effects on fairness perceptions if the subject can provide additional information and ask why a decision was reached, but this also is very much an area for continuing research.

In my view, there is one other element of fairness which society cannot expect the AI developer to implement: the broader aim of achieving wider social equality. Each society's conception of that aim is different, and so any measures which need to be taken by decision-makers ought to be specified through that society's policymaking systems and implemented via law and regulation. This does not mean that an AI user may not choose to attempt to advance social equality through its decision systems. For example, my own university (like many others worldwide) has Equality, Diversity and Inclusion policies which influence its employment decisions and student recruitment. These policies are driven in part by external pressures,[366] but their main source is the internal normative and political decision which the university has made about its role in society. AI producers should note, though, that designing AI decision systems to advance this kind of policy will lead to some unequal treatment, and so the development explanation will need to articulate this policy choice and justify it.

At or post-implementation of the AI, there is far less that AI developers and users can be expected to do to achieve fairness emotions. The most important thing will be to make the development explanation available to its different audiences. Regulators, academics and the AI community might read a detailed explanation early on, but decision subjects will probably not do so unless they have received an adverse decision outcome and are unhappy with it.[367] The development explanation for decision subject might thus usefully concentrate on explaining the steps which have been taken to ensure that adverse decisions are still fair.

A second post-implementation action might be instituting a system of appeal to a human decision-maker. This will not be necessary for most AI decision systems, but ought to be considered in those cases where the consequences of an adverse decision are particularly significant for the subject. If the parallel human decision system incorporates an appeal process, as we find in employment dismissals or criminal sentencing, that is a strong indicator that an AI equivalent should do

[366] See n 69.

[367] For example, researchers have found that consumers almost never read contract terms before purchasing, referring to them only if a dispute subsequently arises: Florencia Marotta-Wurgler, 'Does contract disclosure matter?' (2012) *Journal of Institutional and Theoretical Economics (JITE)/Zeitschrift für die gesamte Staatswissenschaft* 94.

the same. An element of research into decision subject reactions to their decisions might be valuable – if the AI's decisions regularly produces subjects who feel aggrieved by perceived unfairness, a human appeal system would be an obvious way to achieve or restore societal acceptance of the AI system.

The third possibility is providing outcome explanations where necessary, concentrating on those techniques which have been shown to communicate best with decision subjects.[368] Where the AI's decisions form part of a regulated activity, the AI user may be required to provide more detailed outcome explanations to the regulator if some potential regulatory breach is investigated, and the facility to do this will probably need to be designed in by the developer.

Finally, as is good practice for all AI systems, society can expect an appropriate level of monitoring, review and retraining. The main function of this process is to identify decisions which mis-classified decision subjects in terms of the merit criteria, and then to retrain the AI to reach the correct decision in those cases. Additional methods of review and retraining to deal with unfair decisions will also need to be developed and implemented.[369] Fairness is a continuing obligation to society, not a one-time achievement which can then be ticked off as complete.

6.3.2. AI Fairness Regulation to Meet the Social Conception

As a thought experiment, let us imagine that we are the regulator of some activity where humans make decisions about other humans and are under an obligation to do so fairly. Those who conduct that activity tell us that a black box is under development which will be able to replace the human decision-makers – relevant information is fed into the box, and a decision comes out. For reasons of commercial confidentiality we can't be told what is happening inside the black box. As regulator we don't have the power to stop its use, but we do have the power to

[368] One particularly effective technique for explaining to a non-technical audience is by use of counterfactuals. Counterfactual explanations take a decision and explain how it would have differed if some facts had been different – an example might be in a loan application, where a rejected borrower could be told the changes to their income, existing debt, credit history, etc which would have led to the loan being granted: '"explanation" typically refers to an attempt to convey the internal state or logic of an algorithm. In contrast, counterfactuals describe a dependency on the external facts that led to that decision.' Sandra Wachter, Brent Mittelstadt and Chris Russell, 'Counterfactual Explanations Without Opening the Black Box: Automated Decisions and the GDPR' (2018) 31 *Harvard Journal of Law & Technology* 841, 845.

For more detailed discussion see Katja de Vries, 'Transparent Dreams (Are Made of This): Counterfactuals as Transparency Tools in ADM' (2021) 8 *Critical Analysis of Law* 121; Atoosa Kasirzadeh and Andrew Smart, 'The Use and Misuse of Counterfactuals in Ethical Machine Learning' (2021) *Proceedings of the 2021 ACM Conference on Fairness, Accountability, and Transparency* 228.

[369] Jakob Mökander and Luciano Floridi, 'Ethics-based auditing to develop trustworthy AI' (2021) 31 *Minds and Machines* 323.

require it to make fair decisions (or at least to impose regulatory sanctions on those in charge of the activity if decisions are unfair), and we are assured that the black box developers will happily design it so that it meets our regulatory requirement for fairness. How would we proceed?

Our first thought might be simply to extend our existing regulation to cover decisions made by the black box – indeed, it probably already does cover black box decisions indirectly because it imposes a fair decision-making obligation on the humans responsible for the activity. As we noted in chapter five, fairness regulation for human decision-makers usually states no more than that decisions must be fair. The assumption is that fairness is a universally understood human concept, at least within any given society or social grouping, which means that the human decision-makers will understand what they have to do to comply with our regulation.

A little further thought, though, will tell us that this is not a satisfactory solution. The first issue is that perceptions of fairness are based in large part on the justifications for decisions, and the black box merely produces a result with no reasoning or justification. The second is that our regulation gives the black box developers no guidance at all as to what we mean by fairness. This means that it is highly likely that many of the initial decisions made by the black box will be perceived as unfair, which is undesirable in itself because there will be too many adverse effects on decision subjects. Furthermore, time and money will be wasted both in assessing numerous decisions for fairness in the absence of reasons, and in continually redesigning the black box in an attempt to hit an invisible target.

Fortunately we, as the regulator, have a strong understanding of the social conception of fairness, though none about the technology involved. We must therefore try to translate our understanding into some regulatory system which guides the black box developers to a design which produces fair decisions.

The first step must be to unpack the social conception of unfairness into factors which the black box developers might use. From the analysis in the preceding sections, we might come up with the following list:

- Like cases should be decided alike.
- The merit criteria for a positive decision need to be articulated – we will need to decide as regulator how far we specify those criteria, and how far we leave them to those responsible for the activity (the users) to decide.[370] However, all those involved in the regulatory system – the developer, the users, us as regulator and the decision subjects – need to know what these merit criteria are.

[370] Who decides on merit criteria will depend very much on the nature of the activity. If it is selection for medical treatment, we might expect strong involvement from the medical regulator. By contrast, financial regulation rarely sets out criteria for granting loans (except perhaps negative criteria which must not be used), relying on competition in the market to make the widest possible range of lending available.

- In determining whether a decision subject satisfies the merit criteria, some account needs to be taken of any disadvantage which the subject could not have avoided by making different, voluntary, decisions. This will require the development of a regulatory policy on disadvantage, perhaps identifying some elements of disadvantage which the regulation requires to be compensated for and leaving it to users to determine what other disadvantage compensation is appropriate for the activity in question. Again, we would want all these decisions to be articulated.
- Most of the procedural fairness rules we currently apply to human decision-making will not be able to be applied to black box decisions, but we will certainly want decisions to be based on a full set of accurate information. At the outset we may be unsure what a 'full set' of information consists of, unless we can extract that by analysing past human decisions, but experience will gradually build up our knowledge in this area and may lead us to build a definition into our regulation.
- Decisions made by the black box must not unlawfully discriminate against protected groups or individuals, and more broadly should not exhibit any irrational bias (other than the inevitable bias in favour of those who meet the merit criteria). A decision made on a ground which has no apparent causal connection with the outcome is irrationally biased and unfair.

Our second step will be to deal with the lack of reasons for black box decisions. Ideally we would like reasons for each individual decision, but it is highly likely that the black box developers will only be able to offer us reasons which explain the information inputs to the box, and perhaps some technical explanation of how they translate into the output decision. This second explanation will bear no similarity to the kinds of reasons we are used to from human decision-makers.

The alternative is to require the AI user, working with the AI developer, to explain the steps which have been taken to attempt to meet the fairness criteria we have specified, and to justify why those steps are likely to lead to the black box making fair decisions. This explanation will enable us as regulator to assess how well compliance is likely to have been achieved, and we will clearly also want to oblige the developer or user, or perhaps both, to ensure that the black box operates in accordance with this explanation.

Our third step will be to decide whether, when the decision has a substantial negative impact on its subject, there should be some option for the subject to appeal or ask for a review of the decision. This will require us to define the kinds of impact which should trigger this option, and also what form the review should take. At least in the present state of the art, human review is likely to be the only available option. Our decisions on these points will be likely to be driven both by the need to ensure the decision subject is treated fairly and, perhaps most importantly, by the need to secure social acceptance of the system of black box decision-making. For example, society is unlikely to be satisfied that a person accused of a crime should be convicted and punished merely because the black box so decides, but is less likely to complain if lending decisions are made in that way.

Our final step as regulators will be to require systems to be put in place by the user to monitor decisions to identify unfairness, and to ensure that the developer takes steps to modify the black box to deal with those unfairnesses in future decision-making.

In these broad terms, this looks like a regulatory system which should be capable of satisfying society's demand for fair decisions. The question which remains to be answered in the next chapter is whether, if we substitute an AI system for the black box, AI developers and users will be capable of meeting the same requirements.

7

AI Fairness Techniques

To understand how far AI producers might be capable of producing AIs whose decisions accord with the human conception of fairness, to an acceptable level at least, we need to investigate three things. First we need to understand, in very high-level terms, how machine learning is used in AI development. Second, we need to recognise how merit criteria are set and their role in making fair decisions. Finally we need to investigate the various techniques which the AI community has developed in an attempt to achieve AI fairness, and in particular how they attempt to achieve equal treatment and the processes they use to identify and counter bias.

7.1. Machine Learning and Decision Systems

The simplest form of decision system is a decision tree. At each node in the tree it branches, and the information inputs determine which branch the decision follows. The branches lead to further nodes, until ultimately a decision is output.

We could construct an example to decide if a claimant C is entitled to compensation from a defendant D under the common law tort of negligence:

Figure 7.1 Decision tree for the tort of negligence

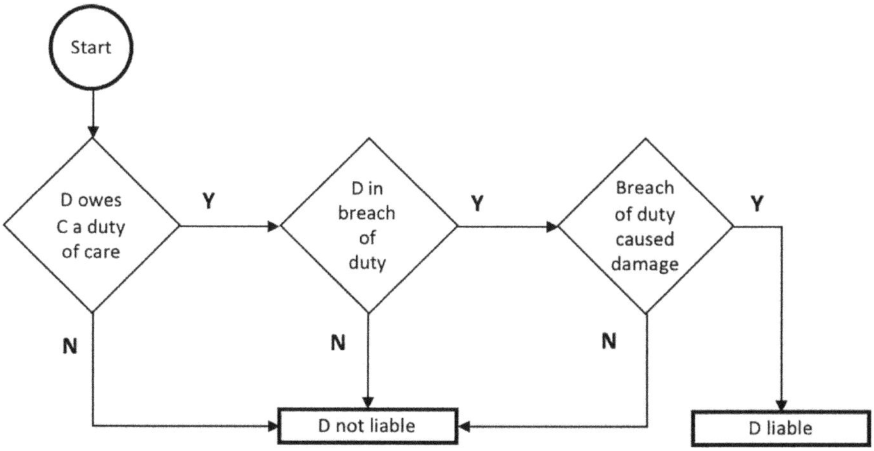

However, as common law lawyers know (and civil lawyers will recognise), Yes/No answers to these questions are not objective data. Instead, a further process of analysis is needed to answer each question. In the early days of expert systems research computer scientists assumed that law was a system of rules, and that therefore decision trees to answer each legal question could be constructed.[371] Lawyers were quick to point out, though, that the meaning of law when applied to a particular case was not simply a matter of semantic analysis of its wording,[372] but rather required an understanding of its meaning *in the context of* that case.[373] In the decision tree in Figure 7.1, any attempt to break down the three questions into sub-decision trees founders when faced with the concept of reasonable foreseeability,[374] which is relevant to each and which cannot be expressed as a decision tree. Legal theorists describe this kind of question as 'open textured':

> All rules involve recognising or classifying particular cases as instances of general terms, and in the case of everything which we are prepared to call a rule it is possible to distinguish clear central cases, where it certainly applies and others where there are reasons for both asserting and denying that it applies. Nothing can eliminate this duality of a core of certainty and a penumbra of doubt when we are engaged in bringing particular situations under general rules. This imparts to all rules a fringe of vagueness or 'open texture'.[375]

Any rule that a person must act fairly is clearly an open-textured rule, requiring human judgement to decide whether that person's conduct falls within or outside the rule.

Many other decision-making domains suffered from the same problem, and so expert systems as a research topic largely faded away, at least in the field of legal decisions. However, the advent of machine learning in recent years has opened up the possibility that a computer might be taught how to replicate the judgement decisions which humans make, by teaching it how to decide using examples of previous decisions by humans: 'ML entails learning patterns and relationships from data to build generalisable models that, when exposed to new or unseen data, assist in a range of general tasks, including categorisation, profiling, prioritisation, filtering and prediction.'[376]

[371] See eg Marek J Sergot, Fariba Sadri, Robert A Kowalski, Frank Kriwaczek, Peter Hammond and H Terese Cory, 'The British Nationality Act as a logic program' (1986) 29 *Communications of the ACM* 370.

[372] These might be the words used in legislation, or the words used by judges to describe common law principles.

[373] See eg Richard E Susskind, 'Some preliminary considerations concerning expert systems in law' (1987) 14 *North Kentucky Law Review* 211, 224–29.

[374] For example, whether D owes a duty of care to C depends on whether D ought reasonably to have foreseen that D's actions or inaction might cause harm to C; *Donoghue v Stevenson* [1932] AC 562.

[375] HLA Hart, *The Concept of Law*, 2nd edn (Oxford, Oxford University Press, 1994) 123, and see also 128–36.

[376] Jatinder Singh, Ian Walden, Jon Crowcroft and Jean Bacon, 'Responsibility & Machine Learning: Part of a Process' (2016) 4, papers.ssrn.com/sol3/papers.cfm?abstractid=2860048.

In simplistic terms, the AI model works through a training database of previous decisions and produces an output for each, effectively a prediction of the decision[377] the human made in that instance. These outputs are compared with the actual human decisions, initially showing little correlation. The model is then reconfigured, either reconfiguring itself autonomously or with guidance from the AI developer, and applied again to the training set. The process is reiterated until the model achieves a high enough success level, which is determined by applying it to a fresh set of test data to see if it achieves a similar or better success rate than humans. The model can then be used on new data to predict what a human would have decided.[378]

This machine learning process creates three difficulties in ensuring that the AI produces fair decisions:

- The AI model will learn to be as fair as its training data, no more and no less. Thus any biases or other unfair aspects of the human decisions in that dataset should be replicated faithfully in the AI. The only way to avoid this difficulty is to eliminate unfair decisions from the dataset, a process which can only be carried out by humans at present, and this is problematic because effective machine learning requires very large training datasets.
- Although the AI model learns how to classify its input data to match or exceed human performance, this does not mean that it performs that task in the same way that humans do. The AI merely finds patterns in the data which correlate with outcomes, and those patterns may have no correlation with how humans determine outcomes. For example, a recruitment selection AI might find a correlation between candidates who have gaps between periods of full-time employment and successful performance in the job role, even though human selectors are unlikely to identify or use that correlation. This is problematic if the correlation is a proxy for some underlying characteristic not included in the data – in our example, the gaps in employment might in many cases be due to pregnancy, raising the potential for unintended unlawful discrimination on the grounds of pregnancy, sex or both.
- Identifying and correcting either of these issues requires some understanding of how the AI is coming to its decisions, and for a human we would achieve this by reviewing retrospectively the facts and reasoning which produced the decision. Unfortunately, AIs which are complicated enough to take over human decision-making are inherently hard to interpret because of that complexity. Retrospective review of AI decisions is likely to require the use of analysis tools which identify the probability that particular data influenced

[377] 'Decision' is used in a very broad sense here – for example, the training data might be a set of images which are either cats/dogs, or more usefully cancerous/non-cancerous organs, and the decision is which category to place the image into.

[378] For an accessible brief description of the different types of AI model involved and their training and testing, see Singh et al (n 376) 4–9.

> the decision, and how much, and from this some explanation of the most likely reasons for the decision can sometimes be constructed. However, this sort of explanation is quite different from the explanations which humans give, which makes it hard to identify any unfairness.[379]

The possibility that unfairness is invisibly embedded in the AI via machine learning, coupled with the difficulty in identifying the AI's 'reasoning' and linking that to fairness, makes retrospective fairness review of AI decisions ineffective. Statistical analysis of the AI's overall decision-making can identify suspicions of unfairness – a loan assessment AI which granted 60 per cent of applications by women but only 40 per cent of applications by men would obviously be suspect – but such analysis cannot say that the AI is definitively being unfair, nor can it identify the causes of any unfairness and how to correct them.

Thus AI developers have concentrated on building fairness into the AI's decision-making from the outset. If we are to impose regulatory obligations of fairness on them, and on the users of their AIs, we need to understand how effective their fairness techniques are in achieving the social conception of fairness which our regulation is aiming at, and to ensure that the regulation places achievable demands on them.

There may be one further issue which is relevant to fairness: AI decision systems trained via machine learning may not be able to adapt to changes in the social conception of fairness. The initial training of such an AI is via a dataset of human decisions, and thus we can assume that the AI will learn at least some aspects of human fairness. From that point on, though, decisions will be made solely by the AI. There will be no new human decisions on which to retrain it, and thus no capacity to evolve the AI's conception of fairness to match changes in society. Even worse, if decisions made by other AIs in the same field are added to its training dataset, there appears to be a danger that retraining might cause the AI to 'forget' nuances of fairness, or even to modify its conception of fairness in unexpected ways.[380]

7.2. Merit Criteria

Where an AI is making a decision about resource allocation which affects human individuals, the whole purpose of using the AI is to choose between those humans so as to decide which should receive the resource. This is done on the basis of some criteria which determine which human merits the resource. To take some of the examples we have referred to throughout, merit could include ability to repay

[379] See Reed, Grieman and Early (n 70) for a detailed discussion of this problem.

[380] This risk has been identified in generative large language models, see Ilia Shumailov, Zakhar Shumaylov, Yiren Zhao, Yarin Gal, Nicolas Papernot and Ross Anderson, 'The Curse of Recursion: Training on Generated Data Makes Models Forget' (2023) *arXiv preprint arXiv:2305.17493*. How far it is a risk for other types of AI is not yet known.

(loan decisions), likelihood to benefit (medical treatment allocation decisions) or suitability for employment (shortlisting decisions). Unequal treatment according to these criteria is what the AI is designed to achieve.

Setting merit criteria for these decisions is not itself unfair according to the social conception of unfairness.[381] However, it can be unfair if the merit criteria also discriminate against those with particular characteristics where the characteristic is not relevant to the aim of decision. An easy example might be using a person's height as one of the merit criteria for a desk job. Height has no correlation with ability to perform the desk job, and so preferring tall people over short (or vice versa) would be perceived as unfair. However, height might be relevant for the proper performance of some jobs – height requirements are common for military, police and firefighting work – and in those cases, using height as a merit criterion will not be unfair.

Most fairness issues with merit criteria arise in relation to apparently rational merit criteria which, quite possibly without the criterion setter realising, place some categories of individual at a disadvantage. An example might be the US case of *Washington v Davis*[382] where employment as a police officer in Washington State required passing a test of verbal ability, and it was shown that black applicants had a failure rate four times higher than white applicants. The US Supreme Court upheld the policy on the grounds that the State's policy was justified by an aim to improve the communication skills of its police force, and that there was no evidence of intent to engage in racial discrimination. Other country's equality laws apply even if there is no intent to discriminate, and so a similar case in the UK would have investigated in more depth the justification behind the test of verbal ability – if a different test would equally well have met the policy aim without producing differential pass rates, the imposition of the original test would be likely to have been held to be unintentional but unlawful discrimination.

This means that those who set merit criteria for AI decision systems ought, on grounds of fairness, to investigate whether the proposed criteria are harder for some groups to fulfil than others. If so, using those criteria needs to be objectively justifiable in terms of what the decision system is intended to achieve. In some cases the justification will be easy. For example, it appears to be harder in the UK for people with a disability to qualify as solicitors than it is for others, as only five per cent of solicitors are disabled compared with 14 per cent of the wider workforce.[383] Even so, if a job requires qualification as a solicitor, setting that as a merit criterion is objectively justifiable and thus not unfair. By contrast, our example of an AI determining who should receive a scarce medical treatment

[381] Though, as noted by Kasy and Abebe (n 44) 576, merit criteria entrench a pre-existing conception of how individuals should be discriminated between. Whether this is unfair at a wider social level is a question which can only be addressed at that social level, and is neither challenged nor answered by the AI community through technical measures.

[382] n 260.

[383] Percentages from the UK Solicitors Regulation Authority, www.sra.org.uk/sra/equality-diversity/diversity-profession/diverse-legal-profession.

is far harder to set merit criteria for. A merit criterion based on quality-adjusted life years (QUALYs) might appear objective, but is clearly likely to exclude older people, those who suffer additionally from another illness, most probably poorer people, and so on. In a complex example like this, fairness in setting merit criteria is most likely to be achieved by adopting a process which society will perceive as fair, which in this case might consist of consultation with medical professionals, patient representatives and relevant civil society organisations and taking their views into account.

Sometimes, though, merit criteria are not set by the AI developer. Instead they are derived via a process of unsupervised machine learning, which analyses a dataset of previous human decisions to identify correlations between combinations of data and results.[384] Those correlations become the merit criteria for future decisions by the AI. The problem here is that it may not be feasible to review those merit criteria as suggested above, because they are embedded invisibly in the AI rather than explicit. Additionally, we have noted at several points previously that a training dataset of human decisions will inevitably contain elements of the bias and unfairness found in humans, so decisions made using the merit criteria identified by the unsupervised learning process are likely to replicate any unfairness in the dataset. This suggests that an important fairness task for AI developers is to guard against this kind of embedded unfairness, either by removing bias from the training set, or identifying and correcting it post training, or both.

All this assumes that merit can be accurately determined, but this assumption is questionable. Take the example of job applicants, where the system for shortlisting them is looking for merit criteria which include qualifications, previous experience, conscientiousness and initiative. Qualifications are matters of objective fact, but experience rather less so – for example, the system will have to decide how to rank five years in a near-identical role compared to seven years working in a similar role, which is a qualitative rather than an objective assessment.[385] Conscientiousness and initiative are not objectively measurable, and can only be induced from matters such as the candidate's record of achievement, statements by referees, etc.

A human decision-maker engaged in shortlisting will be presented with a range of information, some structured via the application form and some selected by the applicant set out in an application letter or a free text 'Additional

[384] For a short explanation of unsupervised learning which is moderately accessible to those from other disciplines, see Happiness Ugochi Dike, Yimin Zhou, Kranthi Kumar Deveerasetty and Qingtian Wu, 'Unsupervised learning based on artificial neural network: A review' (2018) *2018 IEEE International Conference on Cyborg and Bionic Systems (CBS)* 322.

[385] See Carey and Wu (n 44) 14 discussing the difficulty of making accurate comparisons using only predefined criteria. For this particular example, readers might note that time is not really an objective measure of experience. From my 25 years as a glider pilot I learnt that although in the aviation community experience is quantified in terms of flying hours, it is universally accepted that these hours do not have equivalent value. The question is often expressed as being whether the pilot flew 100 different hours, or the same hour 100 times.

Information' section of the form. This at least meets the procedural fairness requirement to hear the decision-subject's case for a positive decision, though in practice there is no way to know how it influenced the decision-maker unless reasons for acceptance or rejection are given.

An AI decision-maker is far more constrained, however. Friedler et al[386] point out that the AI developer must decide in advance which features of the decision subject are to be captured and how they are to be measured. Information about the individual (the 'observed space') is used to create a 'construct space' which is intended to represent the individual in terms of the chosen features which will be used to make the decision.

> In order to define a feature space, we must answer questions about what features should be included and how (and when) they should be measured. This description illuminates our first important distinction from a common set-up of such a problem: the feature space itself is a representation of a chosen set of possibly hidden or unmeasurable constructs. Determining which features should be considered is part of the determination of how the decision should be made; representing those constructs in measurable form is a separate and important step in the process …
>
> In reality, we might not know these features or even the true similarity between individuals. All we have is what we measure or observe.[387]

This means that the AI decision system rests on an implicit assumption that the observed space is identical to the construct space, or more simply that representation of the individual which the AI uses to make the decision is accurately derived from the information about that individual and that it accurately maps to the desired characteristics which determine the decision. The authors describe this as the What You See Is What You Get (WYSIWIG) assumption and give an example from university admissions:

> In the college admissions setting, the WYSIWYG is the assumption that features like SAT scores and high-school GPA (which are observed) correlate well with the applicant's ability to succeed (a property of the construct space). More precisely, it assumes that there is some way to use a combination of these scores to correctly compare true applicant ability.[388]

Adopting this assumption produces decisions which are fair in the sense that equivalent applicants are treated equally, but might be unfair in the sense that the outcomes are biased against particular groups such as those who are disadvantaged.[389]

[386] Sorelle A Friedler, Carlos Scheidegger and Suresh Venkatasubramanian, 'On the (im)possibility of fairness' (2016) *arXiv preprint arXiv:1609.07236*. See also a shortened version: Sorelle A Friedler, Carlos Scheidegger and Suresh Venkatasubramanian, 'The (im)possibility of fairness: Different value systems require different mechanisms for fair decision making' (2021) 64 *Communications of the ACM* 136.

[387] ibid, 3.

[388] ibid, 6.

[389] ibid, 10–12.

This leads us to the two main focuses of AI fairness research:

- achieving equal treatment; and
- countering inherent bias.

As Friedler et al argue, these are inherently incompatible because they represent different understandings of fairness. The AI developer's choices about how to assess merit are themselves fairness decisions, and we will see in section 7.7 that each of these understandings only addresses a part of the broader human conception of fairness.

7.3. Fairness Assumptions

The fairness techniques which have been, and continue to be, developed by the AI research community are based on a number of assumptions about what fairness means. These assumptions are grounded in a Rawlsian view of equality as expanded by later philosophical work – 'Researchers and industry developers in artificial intelligence (AI) and natural language processing (NLP) have uniformly adopted a Rawlsian definition of fairness'[390] – and readers are referred back to section 4.1 for an analysis.

The starting point is that fairness is achieved if all decision subjects are treated equally by the AI, in that they have equal opportunities for a positive decision outcome. However, equality of opportunity comes in three main flavours, as explained by Carey and Wu.[391]

The first is what they describe as formal equality of opportunity. The AI's decisions must not be biased in favour of or against any particular groups, but pre-existing disadvantages are ignored. This is, in effect, the WYSIWIG approach of Friedler et al discussed in the previous section. To achieve this, training data is stripped of characteristics which are irrelevant to the merit criteria of the AI and which might encode pre-existing bias in the data. However, it has been demonstrated that this formal approach fails to achieve its aims, and can even lead to greater bias in outcomes[392] as well as enhancing existing privilege by allocating further opportunities to those already privileged.

The second is described as substantive equality of opportunity (see section 4.1.2), and is based on Rawls' definition of equality of opportunity which is worth repeating here:

[390] Anna Katrine Jørgensen and Anders Søgaard, 'Rawlsian AI fairness loopholes' (2022) *AI and Ethics* 1.

[391] n 117. See also Arneson's very similar taxonomy, discussed at length at text to n 135.

[392] Sam Corbett-Davies, Johann Gaebler, Hamed Nilforoshan, Ravi Shroff and Sharad Goel, 'The measure and mismeasure of fairness' (2023) *arXiv preprint arXiv:1808.00023*.

> Social and economic inequalities are to be arranged so that they are both:
> a) to the greatest benefit of the least advantaged … and
> b) attached to offices and positions open to all under fair equality of opportunity.[393]

Adopting this approach to equality of opportunity requires the AI developer to assume that talent is equally distributed among society, and therefore requires that decision subjects are scored with some weighting to counter any educational or environmental disadvantage which prevents them from demonstrating that talent.[394] Common techniques which aim to achieve substantive equality of opportunity apply weightings so that each group identified for analysis gains an equal proportion of the resources to be allocated.[395] A crude example, far too unsophisticated for actual machine learning fairness use, would be to correct for the previously noted difficulty which disabled persons experience in qualifying as UK solicitors (five per cent as against the population proportion of 14 per cent)[396] by lowering the pass mark for all members of a cohort of disabled entrants to the solicitors profession by an amount which results in 14 per cent of them passing the examination.

The third approach is luck egalitarianism, deriving from the work of Dworkin (see section 4.1.3). Here the AI developer is required to distinguish disadvantage imposed on individuals ('brute luck') from disadvantage which results from their voluntary choices. Scoring is weighted to compensate for the former, but no compensation is applied for the latter. Distinguishing the source of disadvantage is impossible in practice, and so AI fairness techniques tend to adopt Roemer's extension of this idea[397] by substituting effort for choice. Decision subjects are categorised into groups based on a defined set of disadvantages which are generally accepted to arise from brute luck, and then it is assumed that differences between them are the result of exerting greater or less effort. The resultant rankings thus control for luck disadvantage, but make it impossible to compare individuals from different groups to decide which best meet the merit criteria.[398]

It should be obvious that these three assumptions about what fairness consists of are incompatible. AI developers need to choose between them, and that choice is also a normative one, expressing a view about how the world ought to be. As Khan et al write:[399]

> [I]f we view different fairness measures as promoting different conceptions of EOP [equality of opportunity] (formal vs substantive), then this incompatibility is wholly

[393] Rawls (n 110) 302.

[394] Friedler et al (n 386) 8 describe this as the 'We're All Equal' approach, assuming that structural bias exists in the world and correcting for that bias.

[395] See eg Cynthia Dwork, Moritz Hardt, Toniann Pitassi, Omer Reingold and Richard Zemel, 'Fairness through awareness' (2012) *Proceedings of the 3rd innovations in theoretical computer science conference, ACM* 214.

[396] n 383.

[397] For detailed explanation see text to n 140.

[398] See further Falaah Arif Khan, Eleni Manis and Julia Stoyanovich, 'Fairness as equality of opportunity: normative guidance from political philosophy' (2021) *arXiv preprint arXiv:2106.08259* 5–6.

[399] ibid, 6.

unsurprising. From a philosophical perspective, the idea that values trade off against one another is expected.

Consider the EOP-based fairness views that we would expect to be in conflict. Formal EOP rewards people's relevant qualifications, while substantive EOP distributes developmental opportunities equitably. In fair-ML, formal EOP metrics focus on correctly measuring a person's relevant qualifications. Codifications of substantive EOP, on the other hand, either adjust people's qualifications to account for their undeserved circumstances or seek to equalize developmental opportunities at the group level. It is unsurprising that formal and substantive EOP measures would trade off against one another. These views pull in opposite directions, and their corresponding metrics cannot simultaneously be optimized.

We argue that this incompatibility does not impede the practitioner's quest for fair decision-making. Some circumstances call for a more formal approach, while others merit a substantive approach. When it comes to fairness, we must make value-based judgements about which approach – or how much of each approach – is suitable for a given context.

7.4. Fairness as Absence of Bias

Based on these assumptions derived from philosophy, AI fairness research has developed techniques which aim to accord with their chosen fairness assumption. I will describe the resulting algorithmic techniques as 'classifiers', a term commonly used in the literature. This term describes what the algorithm is attempting to do – for the relevant decision (eg shortlist/reject), the classifier calculates a probability, based on the merit criteria, that the decision subject should be placed in the positive group. The highest scoring subjects receive a positive decision, the remainder a negative one. Where scarce resources are being allocated the threshold will be set on that basis, and so might change depending on the resources available.[400] If there is no limit on resources, as might be the case for a loan system where the funds available are not yet exhausted, a fixed threshold will be set and those who exceed it will be offered the resource.

It is worth noting that classifiers are only part of the wider AI decision-system. At one end of the system, the setting of the merit criteria, the selection and labelling of both the training data and data about decision subjects, and so on, need also to be fair. At the other end, any human using the output of the classifier also needs to act fairly. The procedures adopted and built in to the system need to be fair. All these are mainly human decisions, whose fairness we can assess in the normal way, so the research focus in AI is primarily on the fairness of the classifier.

[400] As a simple example, a recruitment exercise will decide to shortlist, say, six applicants, so the top-scoring six will be successful. For a subsequent, identical, post eight might be interviewed, so the threshold there would be different.

All classifiers discriminate between decision subjects – they are biased in favour of some, and against others. University admission systems are biased against those with low educational qualifications. Loan granting systems are biased against those who are unlikely to repay. Ideally, this bias should entirely be based on meeting the merit criteria (suitably adjusted to meet any wider social aims, such as increasing diversity and inclusion). The aim of fairness research is to devise the classifier such that it can be demonstrated that there is no bias in the classifier's decisions which could be attributed to some characteristic of the decision subject which ought not to influence the decision, termed here a 'sensitive' characteristic.[401]

There are numerous possible characteristics which might be tested for bias, so some selection needs to be made. For some possibilities it might seem obvious that they do not influence decision-making and can be rejected.[402] From the others a choice must be made, and this is usually influenced by society's norms about wrongful discrimination. AI researchers have tended to focus on a small number of characteristics which the research community considers to be relevant to unfair bias, and these are (either directly or indirectly) very much influenced by the provisions of national equality laws.[403]

Bias is detected using by statistical metrics to compare the outcomes of the classifier's decisions for those who possess the characteristic chosen for analysis against the outcomes for the remainder of the dataset which does not possess that characteristic. Comparison might be limited to the set of classifier decisions, a purely internal comparison. If the dataset of decision subjects also incorporates the decisions made by humans in their cases, as in a training dataset, then comparison can also be made between the classifier's decisions and those human decisions. A choice also has to be made whether the comparison is based on groups alone, or whether the outcomes of individuals will be compared with the outcomes of other individuals who do not possess the sensitive characteristic.

If the bias which is being tested for is identified, then the classifier is modified in an attempt to reduce the bias and then re-tested. A typical modification might consist of removing some data element from consideration, or weighting some data element or combination of elements, or both. Modification and re-testing continues recursively until the classifier satisfies the test by indicating that the bias has been removed.

[401] 'Protected' characteristic is also used in the literature, but as this is the term used in equality law I prefer 'sensitive' here to avoid confusion. Fairness techniques can be used to test for bias in respect to any chosen characteristic, not just the lists set out in equality laws.

[402] Hair colour in employment decisions might seem an obvious example, but care needs to be taken here. Grey hair is usually a proxy for age. Conventional wisdom is that more women dye their hair blonde than do men, in the UK at least, so a bias towards blonde hair might also be a bias against male applicants. Of course, if hair colour is not captured in the data about decision subject, it cannot influence the classifier's decision and can be eliminated from any fairness analysis.

[403] See Lee et al (n 31) 532–33.

AI researchers have developed a large number of fairness techniques which can be used to check for bias in relation to a selected sensitive characteristic. These fall into five broad categories, each of which is based on different assumptions about what fairness consists of.[404] The first four attempt to measure bias in the classifier's outcomes. The fifth searches for causal links between sensitive characteristics and outcomes.

7.4.1. Internal Consistency

The first two categories analyse the classifier as a standalone system, aiming to assess whether it exhibits bias. The analysis is conducted by dividing the decision subjects into groups – those with the sensitive characteristic and those without – and then comparing outcomes across the groups.

One approach to this task is simply to look for equality of outcome between the two groups, in simple terms that the members of each group have an equal chance of a positive outcome. If this is achieved, it is assumed that the classifier is not biased in relation to that sensitive characteristic. Variations on this theme include statistical parity,[405] equal acceptance rate[406] and benchmarking.[407] This approach provides a broad measure of fairness, but will fail to detect bias if the two groups are not similar in respect of other relevant characteristics. For example, a recruitment selection classifier which selects 50 per cent of all men and 50 per cent of all women for interview would be unbiased on this measure. Suppose, though, that the average educational qualification score for the women in the dataset were higher than the average score for men, and that this score is one of the merit criteria. The social conception of fairness would suggest that the classifier is biased against women, because more of them are better qualified and thus more than 50 per cent should have been selected for interview.

The alternative is to conduct a more granular analysis, based on the probability scores for each decision subject that they will receive a positive (or negative) decision. The aim here is to detect more subtle bias in the classifier, such as the example above where lower scoring men are more likely to be selected than equivalently scoring women. One way to do this is to calculate the average score within each group for those selected for a positive outcome (balance for positive class),

[404] The categorisation and explanation of fairness techniques which follows is drawn primarily from: Verma and Rubin (n 287); Cary and Wu (n 44); Lee et al (n 31) 531–34. Readers should note that this is by no means the only possible categorisation – for just one alternative see Shira Mitchell, Eric Potash, Solon Barocas, Alexander D'Amour and Kristian Lum, 'Algorithmic fairness: Choices, assumptions, and definitions' (2021) 8 *Annual Review of Statistics and Its Application* 141.

[405] Dwork et al (n 395).

[406] Indrė Žliobaitė, 'On the relation between accuracy and fairness in binary classification' (2015) *2nd Workshop on Fairness, Accountability, and Transparency in Machine Learning (FAT/ML 2015)*.

[407] Camelia Simoiu Sam Corbett-Davies and Sharad Goel, 'The problem of infra-marginality in outcome tests for discrimination' (2017) 11 *Annals of Applied Statistics* 1193.

or alternatively to do the same thing for those selected for a negative outcome (balance for negative class).[408] This technique identifies that the classifier is biased in some respect, but does not indicate where the bias lies. A more granular analysis can be conducted based on dividing each group into sub-groups with the same average score and comparing the average outcomes of each sub-group. In our recruitment selection classifier example above we might divide the male/female groups into deciles (scoring 0–10, 11–20 ... 91–100) and analyse for each group, identifying that the classifier was biased against high-scoring women, or in favour of low-scoring men, or perhaps both. Fairness techniques in this category include test-fairness or calibration,[409] matching conditional frequencies[410] and well-calibration.[411]

7.4.2. Correlation with Human Decisions

The third and fourth categories search for bias by comparing the outcomes determined by the classifier with the outcomes of the human decision system which it attempts to replicate. There is, therefore, an underlying assumption here that the human decision system is unbiased, or at least sufficiently unbiased to be socially acceptable.

The third category of fairness techniques compares the two systems as a whole, seeking equivalence in a range of measures: equal chance of a positive or of a negative outcome, equal false positive rates, equal false negative rates, or some combination of these.[412] The analysis of false positives and negatives is particularly useful in detecting subtle biases – returning to our recruitment selection classifier which selects equal numbers of women and men, but where on average women are better qualified by education, this analysis could identify that women are more likely to be wrongly classified as unsuitable, or men wrongly classified as suitable, or both.

The fourth category undertakes a more granular analysis of bias in comparison to the human decision system, by assessing the treatment of comparable individuals from each group. The assumption is that comparable individuals should be treated the same irrespective of the sensitive characteristic, and a straight comparison on this measure is described as a search for causal

[408] Jon M Kleinberg, Sendhil Mullainathan and Manish Raghavan, 'Inherent Trade-Offs in the Fair Determination of Risk Scores' (2017) *8th Innovations in Theoretical Computer Science Conference (ITCS 2017)* 43.

[409] Alexandra Chouldechova, 'Fair prediction with disparate impact: A study of bias in recidivism prediction instruments' (2017) 5 *Big Data* 153.

[410] Moritz Hardt, Eric Price,and Nati Srebro, 'Equality of opportunity in supervised learning '(2016) 29 *Advances in Neural Information Processing Systems.*

[411] Kleinberg et al (n 408).

[412] Describing all these variations would be a distraction here – for more detailed explanation and references see Verma and Rubin (n 287) 3–5.

discrimination.[413] An alternative is to compare the scores for individuals if for each of them the sensitive characteristic is excluded from consideration by the classifier, known as fairness through unawareness – a fair classifier should produce the same result for the comparable individuals from each group.[414] An obvious difficulty here is in deciding whether pairs of individuals are comparable. Finally, fairness through awareness techniques[415] take pairs of individuals, one from each group, and calculate the difference between them based on the merit criteria; the outcomes predicted for each should be a similar distance apart if the classifier is fair. Returning to recruitment selection again, a woman whose match to the merit criteria is 10 per cent better than an equivalent man should have a 10 per cent better chance of being selected. Fairness through awareness avoids the difficulty of identifying comparable individuals because it focuses only on the difference between them, seeking a match between their differences in characteristics and difference in outcome.

7.4.3. Causality

Fairness techniques based on causality[416] take quite a different approach from the previous four categories. They appear to take much of their inspiration from equality law, which prohibits discrimination 'on the grounds of' protected characteristics (see section 5.3.1). In other words, the law prohibits decisions where the protected characteristic *causes* the outcome, in the absence of any legitimate justification for doing so.

Broadly, there are two ways in which causality is identified. The first is to construct a graph[417] which identifies the relationship between those characteristics of the decision subject which are represented in the classifier, and the decision outcome produced by the classifier. The graph also identifies the relationships between those characteristics. An example we might re-use here is the characteristics of sex and height in an employment selection AI. Sex is a legally protected

[413] ie that the outcome (discrimination between individuals) is caused by factors other than the sensitive characteristic. See Sainyam Galhotra, Yuriy Brun and Alexandra Meliou, 'Fairness testing: testing software for discrimination' (2017) *Proceedings of the 2017 11th Joint meeting on foundations of software engineering* 498.

[414] Matt J Kusner, Joshua Loftus, Chris Russell and Ricardo Silva, 'Counterfactual fairness' (2017) *Advances in Neural Information Processing Systems* 30; Jiahao Chen, Nathan Kallus, Xiaojie Mao, Geoffry Svacha and Madeleine Udell, 'Fairness under unawareness: Assessing disparity when protected class is unobserved' (2019) *Proceedings of the Conference on Fairness, Accountability, and Transparency* 339.

[415] Dwork et al (n 395); Kusner et al (n 414).

[416] For a helpful review of causality and machine learning see Bernhard Schölkopf, Francesco Locatello, Stefan Bauer, Nan Rosemary Ke, Nal Kalchbrenner, Anirudh Goyal and Yoshua Bengio, 'Toward causal representation learning' (2021) 109 *Proceedings of the IEEE* 612.

[417] For readers outside the sciences, this is not the kind of graph which represents distributions or equations pictorially, such as a bar chart or line graph. Graph theory is the branch of mathematics which studies relationships between objects – see en.wikipedia.org/wiki/Graphtheory.

characteristic but height is not, and so sex might be chosen as the sensitive characteristic. The classifier is assumed to be unfair if the graph identifies a direct path between sex and the outcome. If the graph identifies a path between height and the outcome, the classifier would also be assumed to be unfair because height is influenced by sex, unless the use of height is assessed by the humans undertaking the analysis as being a causally fair characteristic. This question of fairness is socially determined rather than technical – we noted earlier that jobs such as firefighters or police officers often have minimum height requirements and that this is considered to be fair by society, whereas height is not considered causally relevant for desk jobs. These fairness techniques are usually labelled as fair inference, no unresolved discrimination or no proxy discrimination.[418]

The second route to identifying causality unfairness is via counterfactual techniques.[419] In grossly over-simplified terms, these operate on a trained classifier by taking the training data and, for every individual with the sensitive characteristic, switching that characteristic to a different value. For example, all women in the training dataset might be reassigned the sex value of man. The classifier is then run again on the training dataset and the set of outcomes compared to the original outcomes. If they have changed appreciably, this suggests that the sensitive characteristic is being used unfairly to influence the outcome. These fairness techniques are of course far more complex, because of the likely existence in the dataset of proxies for the sensitive characteristic, but it should at least be apparent from this explanation how these counterfactual techniques are likely to work in principle.

7.4.4. Bias and Equality of Opportunity

In most cases researchers working on these fairness techniques ground their work in terms of Rawlsian equality of opportunity. The assumption is that correcting for bias will compensate for inequality of opportunity, or at least go some way towards doing so. It is therefore worth asking how far these bias-correction fairness techniques might help achieve that aim.

Equality of opportunity is a normative concept; those who have suffered disadvantage should not thereby be denied the same opportunities as those who were not disadvantaged. One variant of the concept holds that this should apply only to disadvantages which were imposed on the sufferer, usually described as 'brute luck', whereas disadvantage arising from the individual's voluntary choices should not be taken into account. These ideas were explored at length in sections 4.1.2 and 4.1.3.

[418] See further Niki Kilbertus, Mateo Rojas Carulla, Giambattista Parascandolo, Moritz Hardt, Dominik Janzing and Bernhard Schölkopf, 'Avoiding discrimination through causal reasoning' (2017) *Advances in Neural Information Processing Systems* 30; Razieh Nabi and Ilya Shpitser, 'Fair inference on outcomes' (2018) 32 *Proceedings of the AAAI Conference on Artificial Intelligence* 1931.

[419] Kusner et al (n 414) seems to be the foundational work here.

It should be apparent from the examination of AI fairness techniques above that they do not try to make any distinction between causes of disadvantage, because these causes are not easily discoverable, and so the implicit assumption is that all those who possess the sensitive characteristic used to test for bias have suffered disadvantage. This has two consequences:

- Individuals with the sensitive characteristic who did not in fact suffer disadvantage will receive more favourable treatment than those who are not members of the disadvantaged group. For example, an individual from a race which is recognised to have lower educational opportunity in that society, but whose parents were wealthy and thus purchased the best possible education, will be scored more highly than an individual from a different race would be even though their educational opportunities were equal.
- The elimination of bias according to these fairness techniques in effect weights the scores of the disadvantaged group so that the average (mean) score is the same as the non-disadvantaged group. As there is variation in disadvantage within the disadvantaged group, the most disadvantaged members will receive less compensation for it than the concept of equality of opportunity demands, simply because the group includes less disadvantaged individuals whose score counts towards the averaging. Additionally, there is a real risk that members of the advantaged group are treated unfairly because like cases in each group are no longer treated alike.[420]

A second assumption, which is probably true but is untestable, is that talent and potential is equally distributed across the population. From this it follows that, had the disadvantage not been suffered, members of the group in question would have probabilities of success which match those of the group without the sensitive characteristic. On this assumption, equalising the success rate between the two groups is an effective way of accounting for the disadvantage and thus providing equality of opportunity. It does not compensate them, as such, for past disadvantage, but it does give group members the opportunity to achieve future advantage based on their own effort and ability, and so largely meets the Rawlsian conception of equality of opportunity.

However, this second assumption only holds where the disadvantage merely *reduces* the likelihood than an individual can meet the merit criteria. For example, a lack of educational opportunity can be overcome through further learning if the chance to do so is offered. However, if the disadvantage makes it *impossible*

[420] This is described as 'levelling down', about which Mittlestadt et al observe:

> When fairness can only be achieved by making everyone worse off … something would appear to have gone wrong in translating the vague concept of fairness into practice. Equality should aim to make people better off, not simply to reduce them to a common level of harm.

Brent Mittelstadt, Sandra Wachter and Chris Russell, 'The Unfairness of Fair Machine Learning: Levelling down and strict egalitarianism by default' (2023) *arXiv preprint arXiv:2302.02404*.

or *highly unlikely* to meet the merit criteria, none of these fairness techniques can achieve equality of opportunity. If we think back to the example of an AI which allocates scarce medical treatment, a particular childhood illness might reduce the chance of a positive reaction to the treatment so much that the AI would never recommend that individual for treatment. Correcting for this bias against those who had suffered the illness would grant the treatment to those who can receive no benefit from it, and thus the AI would fail to achieve its primary objective.

The answer might be that this is not the kind of unfairness which fairness techniques should aim to correct. Rather, it is a question of fairness to be considered when setting the merit criteria – will these criteria in practice exclude some disadvantaged individuals, and if so, is it fair to use these criteria?

Ultimately, we can conclude that a classifier developed using these techniques can be predicted to improve the overall level of equality of opportunity for the group who possess the sensitive characteristics for which bias was tested for and corrected. This is still a broad brush approach to fairness, though, which is inherently incapable of matching the social conception of fairness in more than outline.

The main reason for this is that the characteristics which ultimately determine a classifier's decision – the merit criteria and the decision subject's correspondence with those criteria – are not objective matters of fact which are recorded in the data about decision subjects. Fairness techniques attempt to infer those matters from mathematical computation based on the limited known data about subjects,[421] and thus embed assumptions about both the decision subjects themselves and the characteristics which they are attempting to measure. The truth of those assumptions, and whether they lead to sufficiently accurate measurement, are both unknowable.

Additionally, the assumption discussed above that all those who possess the sensitive characteristic are also disadvantaged to the same extent will inevitably lead to some decisions which fail to match society's understanding of fairness. There will be substantial variation within the group of those with the sensitive characteristic, and thus the classifier will fail in individual cases to provide some individuals with the opportunity they should have received if assessed as individuals, whilst advantaging other individuals in the group when a human decision-maker would not have done so.

This last shortcoming of AI fairness techniques leads us to an important question: is it appropriate to assess fairness by reference to groups, or should AI fairness encompass some degree of individual assessment?

[421] Abigail Z Jacobs and Hanna Wallach, 'Measurement and fairness' (2021) *Proceedings of the 2021 ACM Conference on Fairness, Accountability, And Transparency* 375, 375:

> [C]omputational systems often involve unobservable theoretical constructs – abstractions that describe phenomena of theoretical interest, such as socioeconomic status, teacher effectiveness, and risk of recidivism. Because these constructs cannot be observed, they cannot be measured directly. Instead, they must instead be inferred from measurements of observable properties (and other unobservable theoretical constructs) thought to be related to them.

7.5. Group and Individual Fairness

The problem here arises because AI decision-making systems are, by their nature, abstract models of the human decision-making on which they are trained. A human who makes a recruitment decision, or allocates medical treatment, or decides on a criminal sentence, will be influenced by a far wider range of considerations than can be captured as data for input to the AI.[422] The process of abstraction thus collapses differences between individuals, and leads to their assignment to groups within the modelling. Abstraction seems inevitable if an *ex ante* model of decision-making is to be constructed, and this would be true for human decision-makers if we asked them to undertake this task for their own decision-making – additional factors which those humans only observed during the decision process could then not influence the decision because they were not incorporated into the model.

In recent years AI researchers have debated how far fairness techniques which are trained in relation to groups can also achieve an acceptable level of fairness at the individual level. Dwork et al appear to have initiated the debate, arguing that the similarity metrics used in AI fairness techniques 'may reflect the "best" available approximation as agreed upon by society.'[423] This argument is not uncontested, in particular on the ground that individual and group fairness represent two different conceptions of what fairness means,[424] but Binns contends that this is only potentially true if we can take account of the full context in which the decision is made. Once the decision process is abstracted to construct an AI model, 'at this abstract level, individual and group fairness are not only not in conflict, but are in fact just different ways of reflecting the same set of moral and political concerns.'[425] Thus, although various techniques have been suggested to ensure that individuals with similar characteristics are treated in similar ways, these all fail either because it is not practicable to compare individuals in this way, or because statistical measures of dissimilarity such as fairness through awareness[426] are based on assumptions whose truth cannot be demonstrated.[427] Binns therefore concludes that:

> [W]hile these measures are presented as representing two different kinds of fairness, they can just as easily correspond to the same underlying moral concept. Any conflicts between the two would therefore be artefacts of the possibility that the same underlying concept can be operationalised in different ways.[428]

[422] A recruiter might, for example, consider the personality of an applicant and how that might integrate with the personalities of an existing team. A judge sentencing a convicted criminal will try to assess how far that person acknowledges their guilt and repents.
[423] Dwork et al (n 395).
[424] See eg Jacobs and Wallach (n 421) 383.
[425] Binns (n 300) 515.
[426] See section 7.4.2.
[427] Binns (n 300) 516.
[428] ibid, 517.

This discussion suggests that the differences between group and individual fairness in AI decision-making reduce as additional context is taken into account, which leads us to our next issue.

7.6. The Problem of Context

The social conception of fairness is highly context-dependent, as we saw throughout chapters three to six. Context, as a concept, has both a factual and a normative element. The factual element is the circumstances surrounding the decision which *might* be taken into account, and the normative element is the principles which decide which of those factors *ought* to be taken into account. As an example, the fair sentence for a driver convicted of speeding ought to take into account that an injured person was being driven to hospital, but should ignore the colour of the driver's shirt. The normative element is unfortunately circular – factors ought to be taken into account if it is fair to do so (or perhaps more accurately, would be unfair not to do so), and the decision will be unfair if those factors are not taken into account.

AI decision-making models are devised in advance of the decisions which they make, and so there is also a need to decide in advance which information about data subjects needs to be collected in order to provide as much context as is needed. Only relevant material (in terms of normative fairness) should be considered, so the decision about which data collection will be relevant also needs to be made in advance. This is fundamentally different from human decision-making, because human minds are flexible enough to identify relevant additional context and bring it into consideration at the time of decision-making, something which is (at least at present) beyond the capacity of AI. This means that the 'reality' on which AI decisions are based is a simplified model of the world. The actual world also contains all the context which is omitted from the AI model, and this lack of context can render some decisions unfair.[429]

A fair AI thus needs to capture as much of the relevant context as is possible, and to use as much of it as is feasible as inputs to the decision-making model. This identifies two practical problems: can the data be found; and how much data input can the model cope with? AI developers are forced to make pragmatic decisions here, and any AI decision-making system will inevitably work on less information than is potentially available to a human decision-maker.

[429] See further Jean-Marie John-Mathews, Dominique Cardon and Christine Balagué, 'From reality to world. A critical perspective on AI fairness' (2022) 178 *Journal of Business Ethics* 945, discussing this issue in the context of Boltanski's world/reality distinction – Luc Boltanski, *On Critique* (Cambridge, Polity, 2011). The authors' proposed solution is to modify the AI continuously to capture new context as its relevance is identified (952–53), but this is put forward more as a useful principle than a practical exercise.

This leaves two further, interrelated, problems which are important to any discussion of how AI fairness might be regulated. How will it be decided what data is contextually important and thus should be collected; and who should make that decision?

The 'how?' has to an extent been answered by Lewicki et al,[430] John-Mathews et al[431] and Veale et al.[432] They all agree that identifying the data which is contextually important can only be undertaken by humans who are familiar with that field of decision-making, or are capable of analysing the field to identify these matters. This tells us that we cannot make any general statements about what context needs to be captured – this is completely domain-dependent, and therefore needs to be determined at least at a sectoral level and probably at sub-levels within each sector. As an example, it seems unlikely that the necessary context to make financial sector business decisions fairly[433] can be specified collectively for the full range of different financial services on offer, but it might be possible to specify relevant context for individual services such as loan granting or insurance claim assessment.

Answering *who* should decide what context to include is more difficult. Currently this question is decided by the AI development team, and in most cases their decisions are completely opaque to outsiders to that team. It is likely in many cases that the team has incorporated untested assumptions about what context is important, in part because the cost of investigating that question via interviewing practitioners or analysing their work is high. Decisions to exclude context elements may have been made on the ground that it is computationally too difficult to include them. The important point is that it is not possible to for an external observer to assess the fairness of an AI without knowing these matters.

An opaque, 'black box' AI might be acceptable in domains such as loan granting because of the need to preserve commercial confidentiality, and because there is protection of decision subjects via external legal and regulatory controls which require fair decisions and provide mechanisms for post-decision review of fairness. In other domains, though, external guidance on context will be desirable, perhaps through guidelines from industry bodies or regulators. In domains where fairness is critical, such as criminal sentencing, society might expect external regulation to define a minimum set of contextual factors which *must* be taken into account.[434]

[430] Lewicki et al (n 345) 4 suggesting that the organisation whose decision-making is being modelled should be able to identify these matters.

[431] John-Mathews et al (n 429) 946 suggesting analysis by business ethicists using the techniques developed by Science and Technology Studies (STS) scholars.

[432] Michael Veale, Max Van Kleek and Reuben Binns, 'Fairness and accountability design needs for algorithmic support in high-stakes public sector decision-making' (2018) *Proceedings of the 2018 CHI Conference On Human Factors in Computing Systems* 1, suggesting that contextual factors can be captured through unstructured interviews with practitioners in the field.

[433] Readers may recall that a general duty of fairness towards customers is imposed on UK financial services providers – see section 5.3.3.

[434] See the US decision in *State of Wisconsin v Loomis* (n 363).

Finally, we need to consider the contextual problems which arise from the use of 'foundation' AI models,[435] general purpose AI decision-making tools which are to an extent customisable by users and are made available through an AI-as-a-Service model (AIaaS). Lewicki et al argue that 'it is likely that AIaaS will become *the* primary means of AI proliferation and implementation across many industries and organisations (in contrast with "in-house" AI development).'[436] They point out that the generic nature of AIaaS models makes it practically impossible to deal with fairness issues and relevant context at the level of detail necessary to achieve fairness in practice:

> Each provider's AI services are offered to a range of possible customers, each of which can use the services for widely varying purposes. As such, issues of bias can arise both from the nature of the underlying ML model itself *and* the operating context in which a user employs that service. This means that AIaaS might not only contribute to replicating fairness issues at scale by propagating model's intrinsic biases (or other issues) to the vast number of user applications powered by a given service. Rather, it also raises challenges in terms of both identifying and mitigating fairness issues that can arise from a user's particular application of the service. This is because it can be difficult (if not impossible) to pre-empt all possible purposes for which a service might be used, while users employing these services may not have the knowledge, expertise, ability or access to assess the appropriateness of the service within their context and employ relevant safeguards ... one of the key difficulties in uncovering and mitigating unintended bias in AI systems is that many fairness issues will manifest only in particular contexts, meaning that understanding, let alone accounting for all bias concerns in an effort to create a 'generally applicable' AI service appears infeasible.[437]

Detailed analysis by the authors of a range of commercially available AIaaS offerings identifies exactly the kinds of failings which have been discussed above: implicit and unjustified fairness assumptions; limited choice of fairness techniques; limited or no ability to customise the model to incorporate context; lack of testability; and in general, opaqueness such that the user of the AI has no real chance to identify how fair its decisions really are.

This is exactly the kind of situation where external regulation might seem necessary. The current regulatory trend is to deal with issues of this kind at a sectoral level,[438] but AIaaS is intended to be usable in multiple sectors. Thus regulating for fairness in the AIaaS market will require abandonment of this sectoral regulatory focus in favour of a broader, more generally applicable, approach.

[435] Bommasani et al, 'On the Opportunities and Risks of Foundation Models' (2021) *arXiv:2108.07258*. For an accessible explanation see UK White Paper, *A pro-innovation approach to AI regulation* (2023) paras 86–92, www.gov.uk/government/publications/ai-regulation-a-pro-innovation-approach/white-paper.

[436] Lewicki et al (n 345) 2.

[437] ibid, 4 (references omitted).

[438] See eg UK White Paper (n 435); EU proposed AI Act COM(2021) 206 final, Explanatory Memorandum.

7.7. Meeting the Human Conception of Fairness Through Regulating AI Development

As an overall proposition, it seems accurate to say that AI fairness techniques, both currently and in prospect, are unable fully to capture the human conception of fairness. This is unsurprising – fairness is an emotional response which differs between individuals and societies, is based on a range of views about what equal treatment means, and is heavily influenced by context.

Attempts to explain or, at least to an extent, define fairness all therefore require some level of abstract modelling of the concept. This is what we saw in the work of philosophers and economists, and law and regulation also work on a simplified understanding of fairness although their use of human assessment allows some of the richness of the wider conception to creep back in at that stage. Abstract modelling is also what we find in the AI community, and so it should hardly have been a shock that the fairness techniques reviewed above go only part of the way to satisfying society's need for fair AI decisions.

In particular, AI faces the problem of making comparisons between individuals while taking into account the relevant context in which the decision is made. The more information which is used about individuals to make the comparison, the harder it becomes to make meaningful comparisons. The same is true of context – with enough detail about the context, each decision becomes unique and cannot be compared to other decisions. Defining in advance all the individual characteristics and the context which a human decision-maker might consider relevant is clearly an impossible task.

What this tells us is that a regulatory approach which seeks to define which AI fairness techniques should be implemented for particular applications will fail in its aims. All these techniques simplify and abstract the human understanding of fairness, and thus the decisions made by an AI using them will always fall short of what society expects.

This does not mean that trying to regulate for AI fairness is a hopeless task. Although AI fairness techniques cannot at present *guarantee* fair decisions, and will likely never be capable of doing so, they are certainly capable of *improving* the fairness of AI decision systems.

> [G]iven the many complexities of fairness and its contextual nature, it is generally not possible to fully debias an AI system or to guarantee its fairness; rather, the aim is to *mitigate* fairness-related harms and other unwanted consequences as much as possible.[439]

This is equally true for the regulation of human decision-making, which also risks unfairness and which is what law and regulation seek to mitigate.

[439] Lewicki et al (n 345) 2 (references omitted).

Society needs to accept that AI decision systems cannot be asked to achieve the same fairness as a human decision-maker might, simply because AI fairness techniques have to choose between the various abstractions of fairness and can only operate on a pre-specified dataset for the decision subject. However, we *can* expect that AI producers will make a good faith attempt to achieve fair decisions, and will use reasonable efforts to do so.

If we set out the ideal human decision-making process, it might be conceptualised as follows:

1. Establish rules and principles for making a fair decision – these will specify relevant information about the decision subject which should be considered.
2. Apply those rules and principles to produce a determination.
3. Identify additional relevant information and apply that to review and potentially change the determination.

AI decision-making systems only undertake the first two of these steps. The third would require something akin to AGI, which has not yet been achieved and would raise new fairness issues if it were.

This tells us that AI fairness regulation will need to focus on three matters: the rules and principles for fairness on which the AI is based; the information about the decision subject and the context which the AI will use to make its decision; and how those rules and principles are applied to the information about the decision subject. All these are matters which are currently decided by the AI developer and the AI user, and so it is clear that AI fairness regulation would need to focus on the processes through which the AI is developed, customised for the user and implemented by the user.

What, then, might AI fairness regulation look like? To answer this question we need to identify the areas of mismatch between AI fairness techniques and the human conception of fairness which regulation might usefully address.

The underlying conception of fairness is not at issue. Fairness demands that like cases should be decided alike, but with appropriate compensation for disadvantage in order to ensure that equality of opportunity is achieved. The mismatch here is in the method for comparing cases. In human decision-making this is based on emotion – does it feel unfair to the decision-maker that these two cases should not be treated the same.[440] AI decision-making systems make comparisons on the basis of statistics and probabilities, derived through machine learning analysis of previous human decisions. This mismatch is so fundamental that regulation cannot address it. However, the regulatory aim is not to ensure that decisions are made in the same way in both systems, but rather that the outcomes of both are equally fair under the human conception. This suggests that human review of an AI decision system's outcomes might be a useful requirement. That said, one main utility of AI is producing decisions faster and cheaper than the human equivalent.

[440] See n 43 and accompanying text.

Regulation will therefore need to decide what the appropriate level of review is for different AI decision-making applications. Some might only require minimal, or even no, sampling of decisions once the AI is trained to achieve assurance that it is sufficiently fair. Others might need more extensive human review. In some cases, the possibility to appeal to a human decision-maker will be needed to meet the human conception of fairness. The obvious basis on which to determine this level of review is the extent to which the AI decision system impacts on the fundamental rights and the economic interests of the decision subject.[441] The more serious the consequences to the subject, the greater assurance society will need that the AI makes fair decisions.

Next, regulation needs to consider the decision system's merit criteria. Most human decision systems have expressly articulated merit criteria which are often published. Some AI systems have merit criteria which are articulated by the developer or user, though these are often not published. In other AI systems the merit criteria are identified through machine learning on the basis that the AI achieves a sufficiently close match in its outcomes to those in its training dataset. In these cases, if regulation is concerned with the merit criteria then analysis of the AI will need to be undertaken to make the merit criteria explicit. The choice of merit criteria raises fairness issues – some might be explicitly unfair ('prefer men to women'), while others indirectly lead to unfairness ('no breaks in employment record longer than six months' would eliminate more women than men because of maternity leave). Regulation might reasonably require fair merit criteria, at least where the decision has a serious impact on the decision subject. In some cases a regulator might make some merit criteria compulsory, or forbid the use of others. In other cases it might be sufficient to require those responsible for the AI system, its developers and users, to justify the fairness of the merit criteria.[442] Guidelines for AI developers and users would also be useful here. Who should decide on merit criteria will depend very much on the nature of the activity. If it is selection for medical treatment, we might expect strong involvement from the medical regulator. By contrast, financial regulation rarely sets out criteria for granting loans (except perhaps negative criteria which must not be used), relying on competition in the market to make the widest possible range of lending available.

If the design aims of the AI system are appropriate – equality of treatment based on fair merit criteria – regulatory attention then turns to the process by which decision subjects are assessed. A fair decision system avoids unfair bias, which needs to be avoided at two levels:

- Unlawful discrimination is clearly unfair. It will exist where the AI system produces less favourable outcomes for those who have a particular sensi-

[441] The legal conception of fairness is heavily influenced by a decision's potential impact on these matters – see section 5.4.1.

[442] An example thought experiment in relation to loan granting AI is discussed in Hana Chockler and Joseph Y Halpern, 'On testing for discrimination using causal models' (2022) 36 *Proceedings of the AAAI Conference on Artificial Intelligence* 5548.

tive characteristic which is also protected by law. Bias of this kind tends to be generated via the machine learning process, replicating the (usually) unintentional biases of the human decision-makers represented in the training dataset. Equality law is helpful here because it provides AI developers with a list of sensitive characteristics which must be analysed for bias, derived from the law of the jurisdiction where the AI system will be used.[443] The fairness techniques examined in section 7.4 all to some extent identify and correct for such biases, but we noted that they do so in subtly different ways and produce different results. We also recognised the problem of proxies for sensitive characteristics – for example both height and income have some correlation with sex – and AI producers will need to search out and correct for these too. Ultimately it will be for the courts to decide whether particular bias correction techniques are sufficient to avoid unlawful discrimination, though guidelines from the equality regulator (where one exists) would be both helpful to AI producers and useful for judges.

- Discrimination on the basis of some characteristic which has no causal connection to the outcome, in the sense that it has no relevance to the merit criteria, might also be considered unfair. Private entities usually[444] have the legal freedom to act capriciously and exhibit bias, outside equality law, but public bodies may be more constrained.[445] Even where such bias might be lawful – such as a business owner who has an irrational preference for employees with blue eyes[446] – outcomes based on it will be seen by wider society as unfair. This might lead to social pressure to constrain unfair biases of this kind through regulation, which would also accord with the policy of many governments that the social acceptance of AI is to be encouraged.

As we saw in section 7.4.4, bias correction is assumed by the AI research community to provide adequate correction for inequality of opportunity, provided the appropriate sensitive characteristics connected with disadvantage are identified and used. I suggested there that this is only true at the collective level of the group possessing the sensitive characteristic, and that some individuals may still fail to be appropriately compensated for disadvantage while other are over-compensated. Additionally, all disadvantage is assumed to be the consequence of brute luck, so no allowance is made for disadvantage arising from individual choices. The human conception of fairness requires compensation for disadvantage at the individual

[443] AI developers whose AI will be used in multiple jurisdictions, and thus particularly those who produce foundation models offered as AIaaS, might be well-advised to use a composite list compiled from the major jurisdictions where they expect customers.

[444] But not always. Note, for example, the broad fairness obligations imposed in the UK by data protection and financial services law – see sections 5.3.2 and 5.3.3.

[445] See the discussion of procedural fairness and UK judicial review in section 5.1.1.

[446] And noting that blue eyes are potentially a proxy for race, a protected characteristic – see Amanda Pneuman, Zoran M Budimlija, Theresa Caragine, Mechthild Prinz and Elisa Wurmbach, 'Verification of eye and skin color predictors in various populations' (2012) 14 *Legal Medicine* 78.

level, but often considers disadvantage caused by voluntary choices not to be something which should be compensated for.

This mismatch seems inevitable, because of the practical impossibility of making meaningful comparisons at an individual level and capturing data on individual choices. It may be that this is acceptable to society, because the human conception of fairness is complex and nuanced. If a decision-maker *attempts* to act fairly, society might be content with outcomes which are unfair to the individual concerned – for example, if a job applicant is rejected because of poor qualifications, resulting from earlier disadvantage, society might accept both that the employer was fair to make this decision, and also that although it is unfair that the individual lacked the opportunity this unfairness is not something the employer should be expected to correct. It seems unlikely to me that regulation could demand much more from AI producers, particularly as it does not normally demand that human decision-makers achieve equality of opportunity. If, though, a society makes a policy decision to encourage equality of opportunity,[447] regulatory guidance on how AI producers might best help to achieve that would be useful.

Procedural fairness in human decision-making is important, in part because fair procedures are more visible to society and thus increase belief that less visible outcome fairness is likely to be achieved (see the discussion in sections 3.2 and 5.1). Procedural fairness has two elements: lack of bias, which is what AI fairness techniques aim to achieve and is thus built in to the AI; and a fair hearing for the decision subject. Most AI decision systems will not engage in anything like a formal hearing, so we must focus on the purpose behind such a hearing. This is to make sure that the decision-maker is aware of, and thus can consider, all the information which is relevant to the decision. Thus the equivalent standard of fairness for AI would seem to be that its decisions should be based on a full set of accurate information. At the development stage there will be inevitable uncertainty about what a 'full set' of relevant information consists of, but the machine learning process might extract a minimum set by analysing past human decisions. If the AI's performance is monitored to detect potentially unfair decision, experience will gradually build up knowledge in this area and identify further information which is needed for a fair decision.

This takes us to the issue of context, which plays an important part in human decision-making. Human minds are sufficiently flexible to identify differences in context, and these differences change the decision-maker's emotional response to the possible outcomes, identifying which are fairer. Context is the set of information which the decision-maker has available for consideration, and for human decisions there is no limit on what might be considered. By contrast, AI decision systems are more limited, and thus less likely to produce fair decisions in

[447] For an example see UK university admissions policy, OFS Annual Review 2019 (n 69).

all contexts.[448] AI producers define in advance what information will constitute input data for the AI (generative AI like ChatGPT is an exception, but generative AI is very far from suitable for decision systems because it aims at plausibility, not truth and accuracy).[449] Dolata et al argue that this contextual limitation is based on an erroneous assumption that AI fairness techniques are context-independent, leading to the adoption of metrics which induce 'the illusion of mathematically expressing abstract notions without either losses or unexpected consequences.'[450] This creates a mismatch with human fairness – ideally the AI would consider as much context as could be captured, but this in turn creates the risk that its training via machine learning might base decisions on data which a human would consider irrelevant, rendering the decision unfair.

This suggests that regulation might play some role in specifying a minimum set of relevant information for a particular domain of AI application, thus also defining in part the context which should influence decisions. It might also specify irrelevant information, or information which should not be considered because it creates a risk of bias or inaccuracy.[451]

The final area of mismatch which we need to consider is the role which explanations play in achieving fairness. We saw in section 6.3 that explanation are both proxies for fairness – a convincing explanation suggests that the decision outcome is likely to have been fair – and also that they help engender the emotion of fairness in observers and decision subjects. This is true for ex post explanations of individuals decisions and also for advance explanations of how the decision system operates to achieve fairness. Our analysis of the legal conception in section 5.4 identified that the law tends only to demand individual outcome explanations where the decision subject's rights and interests are seriously impacted, but that the ability to provide such an explanation is useful evidence if there is some other ground on which the law might intervene (an example might be breach of a contractual term requiring one party to act with reasonable care, where unfairness of an AI's decision-making might suggest it was not reasonable to use the AI to discharge that obligation).

I have suggested elsewhere that providing explanations about how and why an AI reached its decision is problematic because there is a mismatch between what can be explained about an AI and what humans expect from explanations:

> Humans explain themselves in definitive terms – this is how it happened, this is how I will decide – but AIs are predicated on uncertainty and only tell us probabilities – this

[448] See Veale et al (n 432); Binns (n 300) 518–19.

[449] Alejo Jose G Sison, Marco Tulio Daza, Roberto Gozalo-Brizuela and Eduardo C Garrido-Merchán, ChatGPT: 'More than a Weapon of Mass Deception, Ethical challenges and responses from the Human-Centered Artificial Intelligence (HCAI) perspective' (2023) *arXiv preprint arXiv:2304.11215*.

[450] Mateusz Dolata, Stefan Feuerriegel and Gerhard Schwabe, 'A sociotechnical view of algorithmic fairness' (2022) 32 *Information Systems Journal* 754, 762.

[451] See Chockler and Halpern (n 442), discussing how a financial services regulator might mandate or forbid the consideration of particular information in loan granting AI decisions.

> is most likely to be how it happened, this is probably how future decisions will be made. XAI-assisted explanations for what has already happened might … be similar enough to human explanations once their probabilistic nature is understood. Explanations about the future decisions an AI might make are, though, very different from those about human decision-making.[452]

Humans understand the world best in terms of stories, and so convincing explanations about AI need to be translated from probabilities and metrics into narrative terms. This is potentially achievable, though requiring some human effort,[453] to justify the fairness of an individual AI decision.

Explaining the overall fairness of an AI in advance is more difficult. The narrative which explains the outcome decision for subject A might be quite different from the narrative for subject B. An overall, composite explanation about the AI's decision-making, couched in narrative terms, might be so detailed and complicated as to be incomprehensible to most.[454] Regulating to demand such an overall, *ex ante*, explanation of fairness is likely to make compliance impossible to achieve for some AI producers, and unlikely to achieve the regulatory aim of assuring society that the AI is likely to make fair decisions.

Fortunately, there *is* a kind of narrative explanation which AI producers could provide. They could explain the steps they have taken in their attempt to ensure that the AI makes fair decisions, and justify their choices as to the technical methods they have adopted. Such an explanation might be equally difficult for wider society to understand, but it would have two clear benefits:

- As we saw earlier in this book,[455] the process of devising an explanation also generates a normative impulse to comply with that explanation. Thus the mere fact that AI producers explain their fairness methods and choices increases the chance that they will construct and test the AI to ensure that it achieves their fairness aims.
- Explanations greatly assist external evaluation of AI fairness. They provide a standard against which the actual performance of the AI can be tested. Such explanations will be of particular use to regulators and courts, who will need to decide whether AI producers are likely to, or have in fact, lived up to their legal and regulatory obligations. Additionally, where AI producers are subject to legal and regulatory obligations, constructing an explanation helps them to identify whether they have made a convincing case that they have complied with those obligations.

[452] Reed, Grieman and Early (n 70).

[453] Or perhaps, in the near future, the assistance of generative AI.

[454] Understanding such explanations seems to require at least level 3 literacy on the OECD scale from its 2013 OECD Skills Outlook (www.oecd.org/skills/piaac/Skills%20volume%201%20(eng)--full%20v12--eBook%20(04%2011%202013).pdf) 64. Roughly half the population in the countries surveyed were at level 2 or below, ibid, 63.

[455] Peters (n 167) and accompanying text.

Transparency is often called for as a suitable regulatory tool,[456] though those calling for it often have little idea what it might look like in practice.[457] Although transparency in terms of a detailed description of how the AI makes decisions is an unrealistic demand, which is probably of little utility in any event,[458] realistic and useful regulatory requirements for transparency are possible.[459] Transparency in the form of the kind of explanation discussed above is both achievable for AI producers, and has clear regulatory utility. The need for explanation is also recognised by the AI research community. As Jacobs and Wallach write:

> The contested nature of fairness makes it inherently hard to measure: If there are multiple theoretical understandings of a construct, then it is imperative to articulate which understanding is being operationalized. A failure to do this makes it difficult to meaningfully compare different operationalizations because they may be operationalizing different theoretical understandings. In turn, this makes it difficult to identify mismatches between fairness as it is theoretically understood and any given operationalization … We argue that debates about values should be explicitly treated as such instead of being couched indirectly in mathematics, where they are rendered less accessible to the stakeholders that are most likely to be affected by the computational systems in question.[460]

However, imposing an explanatory obligation on all AI producers is too simplistic. Some AIs need no fairness explanation. For others, a high-level explanation which is meaningful to broader society would be useful. AI decision systems which can have a serious impact on the rights and interests of decision subjects may require more, and if they operate in a regulated sector then the regulator might establish detailed requirements about what needs to be explained. Imposition of this obligation needs to be granular, not universal, and the content required should reflect the risks which the AI decision system generates. These are important considerations for any system of regulation for AI fairness.

[456] See eg US White Paper, Executive Office of the President, *Artificial intelligence, automation and the economy* (2016); British Academy & Royal Society, *Data management and use: governance in the 21st century* (2017); UK House of Lords Artificial Intelligence Committee, *AI in the UK: ready, willing and able?* (2018) HL Paper 100.

[457] Weller identifies eight different types of transparency, each of which requires different measures to achieve and needs to be explained in different ways – Adrian Weller, 'Transparency: motivations and challenges' in Wojciech Samek et al (eds), *Explainable AI: interpreting, explaining and visualizing deep learning* (Cham, Springer International Publishing, 2019) 23, 25. For a critical analysis of current approaches to transparency in the AI community see Eric Corbett and Emily Denton, 'Interrogating the T in FAccT' (2023) *Proceedings of the 2023 ACM Conference on Fairness, Accountability, and Transparency* 1624.

[458] Chris Reed, 'How should we regulate artificial intelligence?' (2018) *Phil Trans R Soc A* 376.

[459] Thomas Wischmeyer, 'Artificial intelligence and transparency: opening the black box' in Thomas Wischmeyer and Timo Rademacher (eds), *Regulating artificial intelligence* (Cham, Springer Nature, 2020) 75–101.

[460] Jacobs and Wallach (n 421) 376.

8

A Regulatory Framework for AI Fairness

I hope that the previous chapters have convinced the reader that society will in at least some cases want to regulate the fairness of AI decision systems. The social imperative that decisions should be made fairly is extremely strong, perhaps even part of the fundamental glue which binds individuals into a society.

Even if there is a clear desire for some kind of AI fairness regulation, this does not tell us how to approach devising that regulation. We have seen that the various disciplines which take an interest in AI fairness have quite different attitudes to the problem.

> [D]ifferent disciplines also embody different research cultures that can affect the clarification and refining of ambiguous concepts. For instance, many machine learning researchers would naturally seek to construct a mathematically precise definition of, say, 'fairness', whereas qualitative social scientists would often seek to highlight the rich differences in how different stakeholders understand the concept. Similarly, philosophical ethicists often seek to highlight inherent dilemmas and in principle problems for different definitions of a concept, whereas many lawyers and researchers from other policy-oriented disciplines would look for operational definitions that are good enough to resolve in-practice problems.[461]

Regulation, as a legal project, needs to aim at pragmatic solutions to real-world problems while avoiding the imposition of burdens which aim to counter future problems that are merely speculative, but this should not mean that regulation is produced in a legal silo. The purpose of AI fairness regulation would be to ensure that, so far as possible, AI implementations satisfy the human need for fair decisions. This requires any regulatory proposal to take account of the insights into fairness from other disciplines, and align with their findings so far as is practically possible.

[461] Whittlestone et al (n 98) 15.

8.1. Regulatory Approaches

Floridi et al have analysed a broad range of AI projects, from which they identify seven factors that they describe as essential if AI is to achieve social good:

> (1) falsifiability and incremental deployment; (2) safeguards against the manipulation of predictors; (3) receiver-contextualised intervention; (4) receiver-contextualised explanation and transparent purposes; (5) privacy protection and data subject consent; (6) situational fairness; and (7) human-friendly semanticisation.[462]

In the context of fairness, 'falsifiabiity' demands that the AI's design should allow its fundamental assumptions, in our case about the effectiveness of the chosen fairness techniques, to be tested. The authors note that we cannot test for truth, because this requires testing for every possible future decision subject, but that it must be possible to test the assumptions across a sufficiently representative range of decisions that we are reasonably satisfied that the assumptions have not been shown to be false.[463] 'Receiver-contextualised explanation and transparent purposes' requires that the purposes for which the AI is being used, and how it makes its decisions, should be understandable by others. This explanation needs to be contextualised for the person who receives it – for example, regulators require more detail, decision subjects need a much more generalised explanation, while engineers might want technical detail.[464] The authors explain 'situational fairness' in terms of bias, with particular attention required where proxies are used which might disguise bias.

These principles, expanding the meaning of fairness beyond bias, seem to accord to a large degree with the findings in this book. Thus they might be a useful set of characteristics which we should look for in any proposal to regulate for AI fairness. We must recognise, though, that the full human conception is broader and we should not confine our regulatory attention to these principles alone.

Fairness in its full human conception is our regulatory ideal, though the preceding chapters have shown that we are unlikely to achieve it completely.

8.1.1. Regulatory Principles

At the legislative level, AI fairness is at present mainly put forward as an ethical principle with which it is hoped that AI producers will comply, but without any regulatory system to incentivise them to do so. The UK AI White Paper includes fairness as one of its overriding principles, but proposes that most AI implementations should be influenced by ethical guidance, with detailed AI fairness requirements (if any) left to sectoral regulation.[465]

[462] Floridi et al (n 44) 1775.
[463] ibid, 1776.
[464] ibid, 1781–84.
[465] AI White Paper (n 435) paras 10–11.

The EU AI Act[466] focuses on risks to health, safety or fundamental rights, and fairness is only a consideration for the fundamental rights to data protection, to a fair trial and to fair and just working conditions.

Perhaps the closest at present to a regulatory system for AI fairness is that of Singapore's Model Artificial Intelligence Governance Framework.[467] This is structured as a model for how Singapore's AI developers and users should behave when developing and implementing AI, and for the governance structures they should adopt to ensure that their AI activities follow the model. It is not drafted as regulation, and sets out no consequences for non-compliance. However, the close integration of Singapore's Government and regulators with the business community means that the Framework is likely to be treated as something very close to regulation, and compliance is highly likely.

The Framework is constructed around two fundamental principles, the first of which is: 'Organisations using AI in decision-making should ensure that the decision-making process is *explainable*, *transparent* and *fair*.'[468] Useful and practical illustrations are given, particularly relating to ensuring that unintended bias is excluded from the design and training of AI systems, and is tested for and corrected if found. However, fairness is still presented as a broad, human-understood concept which the Framework does not seek to define, and where mechanisms to achieve it are left to AI producers to determine.

Most commentators who discuss how AI fairness might be achieved through regulation also concentrate on the principles which regulation ought to implement, with a strong focus on the human rights aspects of fairness. Thus, for example, Cath et al recommend that regulation of AI should aim at promoting the public good, defining the good AI society in terms of the concept of human dignity derived from the EU GDPR, the 1948 Universal Declaration of Human Rights and the EU Charter of Fundamental Rights.[469] This is clearly a laudable regulatory aim, but how to achieve it is left an open question.

Salgado-Criado and Celia Fernández-Aller suggest that the answer might lie in attempting to define sub-types of AI systems and to describe their technology and architecture:

> [J]ust to say that an AI system needs to be trustworthy, fair, or transparent is not enough and leaves it very open to interpretation and implementation, becoming

[466] Regulation (EU) 2024 of the European Parliament and of the Council laying down harmonised rules on artificial intelligence (Artificial Intelligence Act) This book went to press just before the AI Act was published in the Official Journal. The final text adopted on 21 May 2024 is at data.consilium.europa.eu/doc/document/PE-24-2024-INIT/en/pdf.

[467] *Singapore Model Artificial Intelligence Governance Framework*, 2nd edn (2020), www.pdpc.gov.sg/-/media/files/pdpc/pdf-files/resource-for-organisation/ai/sgmodelaigovframework2.pdf.

[468] ibid, para 2.7. There is a recognition that this is a standard to guide the humans who construct and implement AI, and thus not an objective AI performance metric: 'Although perfect explainability, transparency and fairness are impossible to attain, organisations should strive to ensure that their use or application of AI is undertaken in a manner that reflects the objectives of these principles as far as possible. This helps build trust and confidence in AI.'

[469] Corinne Cath, Sandra Wachter, Brent Mittelstadt, Mariarosaria Taddeo and Luciano Floridi, 'Artificial intelligence and the 'good society': the US, EU, and UK approach' (2018) 24 *Science and Engineering Ethics* 505, 524–25.

> useless in terms of enforcement. We need a more formal definition and a technical and architectural description of each type of AI system. An image classifier, a voice or face recognition, an ML that predicts human behavior in the future, or a recommender system: all these groups of applications have already a more or less stable architecture and for each of them we could already start to identify measures of performance and checkpoints that should be tested before and during the algorithm operations.[470]

They propose five checkpoints[471] where regulation might be applied to those systems, but do not offer suggestions about how regulation might actually achieve effective checks on them. Readers might share my serious doubts about how far it is possible for regulation to partition AI systems into defined types and describe them in a way which is both accurate and future-resistant. Any regulatory action of this kind inevitably risks ossifying technological development because of the uncertainty it creates about technology which falls outside a regulatory classification but performs similar functions to technology within the classification.[472]

An alternative to devising new regulation has been suggested by Wachter et al.[473] They analyse the incompatibilities between AI development approaches to fairness and the requirements under EU law that personal data should be collected and processed fairly, focusing on discrimination. To reduce these incompatibilities they propose changes to judicial practice when deciding cases under EU data protection law. The current judicial method of assessment is, in essence, for judges to apply their judicial intuition[474] to the evidence, and to do so in a broad and undefined evidential context in which any evidence which the judge considers relevant can be considered. Instead, the authors suggest that the practice of judges should be modified in two ways: first, by defining and establishing procedures which all judges should use in assessing discrimination, reducing the reliance on

[470] Jesús Salgado-Criado and Celia Fernández-Aller, 'A wide human-rights approach to artificial intelligence regulation in Europe' (2021) 40 *IEEE Technology and Society Magazine* 55, 59.

[471] Those checkpoints accord closely with the elements of fairness identified in the preceding chapters of this book and are, in summary:

- success criteria ('merit criteria' in the terminology I have used in this book in relation to fairness);
- decision process – how the AI arrives at its decision;
- decision itself – the decision outcome in the terminology used in this book;
- effects on people; and
- protection of data.

[472] An example might be the EU Electronic Signatures Directive, Directive 1999/93/EC on a Community framework for electronic signatures, OJ L 13/12, 19 January 2000. Biometric signature techniques fell outside the definitions in the Directive, thus creating uncertainty whether biometric signatures would be recognised as having any legal validity and quite likely slowing further development and commercialisation of the technology.

[473] Sandra Wachter, Brent Mittelstadt and Chris Russell, 'Why fairness cannot be automated: Bridging the gap between EU non-discrimination law and AI' (2021) 41 *Computer Law & Security Review* 105567.

[474] We might equate this to an evaluation of the judge's internal emotions of fairness, which we have noted throughout this book as being core to the human conception, albeit an evaluation by a trained professional who is required to identify and articulate justifications for that emotion from the evidence considered in the case.

intuition; and second, by establishing standard forms of statistical evidence to be used for that purpose.[475]

This proposal might enhance judicial assessment of AI fairness, and over time might feed back into AI development and implementation in ways which improve the overall fairness of those AI systems. However, this will be a slow process. Further, the standardisation of evidence seems likely to require different standards for different types of AI decision-making, and thus might become increasingly complex and granular over time. An additional risk is that this approach ends up by applying different fairness standards to AI and human decisions – I have argued elsewhere[476] that there is a strong normative value in ensuring that the regulation of equivalent activities produces functionally equivalent effects, even if different processes and rules are required to achieve this equivalence – and it is unclear how that risk could be managed if this approach is adopted.

Identifying the principles for AI fairness regulation is important, but not sufficient. As Black and Murray note:

> [M]uch of the discussion on AI has related to the role of goals and values, discussed in the language of ethics. But identifying, and agreeing, goals and values is only one element of a regulatory system: essential, but by no means sufficient. Regulation also requires people and organisations to change their behaviours.[477]

Changing judicial practice can only address a small proportion of the legal fairness demands explained in section 5.3, let alone the broader fairness demands of the human conception of fairness. It seems clear to me that it is necessary to find a regulatory structure which might address the fairness problem, and devise a suitable regulatory scheme which uses that structure.

8.1.2. Regulatory Structures

There is a substantial body of work on regulatory structures, identifying the strengths and weaknesses of each and their suitability for particular regulatory purposes. One of the simplest structures (though not simple to implement) is that proposed by Lessig,[478] where regulation is undertaken directly by code and the

[475] ibid, 4:

> AI does not discriminate in an equivalent way to humans which disrupts established methods for detecting, investigating, and preventing discrimination. To contend with automated discrimination, we encourage the judiciary and industry to move away from predominantly using measures based on intuition, and towards a more coherent and consistent set of assessment procedures (not consistent interpretation) for automated discrimination … [and] Establishing a standard set of statistical evidence for automated discrimination cases can help ensure consistent procedures for assessment, but not judicial interpretation, of cases involving AI and automated systems.

[476] Reed (n 364) 95–108.

[477] Julia Black and Andrew Murray, 'Regulating AI and machine learning: setting the regulatory agenda' (2019) 3 *European Journal of Law and Technology* 11.

[478] Lawrence Lessig, *Code and other laws of cyberspace* (New York, Basic Books, 1999).

role of law is just to mandate the effects which that code must achieve. Gervais[479] advocates this as a regulatory structure for AI, identifying a range of normative principles which he suggests that coders should be legally obliged to implement in AI developments. This form of regulation might at first sight seem attractive to AI producers because it appears to regulate them in terms which they already understand. However, it has two drawbacks which in my view are fatal. The first is that there is likely to be a range of national law demands from each state, which are unlikely to be coordinated and may well be contradictory. This forces AI producers to choose which of those regulatory demands to implement in their code, and thus creates the risk of regulatory enforcement by states whose law has lost out in the normative competition for their attention.[480] More seriously, it assumes that AI developers in particular will be able to understand how to translate regulatory demands into code. We saw in chapter seven that there is internal disagreement within the AI community about the best ways to achieve fair AI decision-making, and that the techniques which they have developed miss important elements of the legal and human conceptions of fairness. A further objection, though not necessarily fatal, is that this form of regulation can over time erode the democratic legitimacy of the regulation as coders gradually determine more and more of its policy.[481]

A second, equally simple, structure is to use the existing law, extending its scope as fairness issues arise and introducing new laws and regulations in reaction to problems which the existing law is shown not to provide adequate solutions for. Viljanen and Parviainen[482] have analysed how the existing laws of Finland, primarily those which implement EU law, map to the issues raised by AI applications. They find that AI is already highly regulated by existing laws, which apply to many aspects of AI systems, but that these legal rules drastically differ in their character, makeup and targets, have varied pedigrees, and emanate from a wide variety of institutional actors. As a consequence, some make adequate provision to regulate AI risks, while the application of others is far from certain and some risks are not covered at all.

> [I]t is important to bear in mind that the regulatory layers in the stratigraphy are internally uneven. This will cause the regulatory intensity varying significantly across application domains. Some application domains are subject to intense regulation. Health care AI, for example, is already subject to intensive data regulation,

[479] Daniel J Gervais, 'Towards an effective transnational regulation of AI' (2021) *AI & Society* 1.

[480] See also Reed and Murray (n 23) for the wider drawbacks of overlapping and contradictory regulatory claims.

[481] Eva Micheler and Anna Whaley, 'Regulatory technology: Replacing law with computer code' (2020) 21 *European Business Organization Law Review* 349, 362: 'Allowing private providers wide discretion in making design choices for regulatory technology could lead to an unwarranted "outsourcing" or even "privatisation" of the regulatory process, creating the risk of undermining sovereignty, and positioning it in a grey zone from a constitutional law perspective.'

[482] Mika Viljanen, and Henni Parviainen, 'AI applications and regulation: Mapping the regulatory strata' (2022) 3 *Frontiers in Computer Science*, article 779957.

> application-specific AI regulation initiatives are underway, and the sector is generally intensely regulated. Other domains may simultaneously remain sparsely regulated or completely unregulated. Some industrial AI applications face no regulation apart from the general tort and criminal liability rules, IPR rules and competition law rules. Even these general non-AI rules might be barely relevant in some AI applications. The unevenness advocates for caution in regulatory mappings. Relying on insights and experiences from a bordering domain could provide inadequate guidance.[483]

It should be unsurprising that the existing fairness laws (see chapter five) do not adequately guide and control developers in achieving AI fairness (see chapter seven) because they were designed and developed in a world which had no significant AI implementations.

Thus we need to find a regulatory model which has the potential to be the basis for an AI regulatory system. There are plenty to choose from.[484]

8.1.2.1. Command and Control

We can reject at the outset what we might describe as the 'standard' or command and control approach of setting up a regulator who tells regulatees in detail how to behave in order to achieve the regulatory aim. This model will not work for AI fairness for two main reasons.

First, we have seen in the preceding chapters that a definition of fairness which might be useful for regulatory purposes remains elusive. Technological research into AI fairness has, understandably, focused on quantitative techniques for comparing differences between decision subjects, and has consequently made little progress towards translating the wider human conception of fairness into computable metrics whose use regulation can mandate.[485] Producing the necessary metrics may even be impossible because fairness is an emotion, and so command and control regulation would, in effect, be prescribing behaviour which is intended to achieve that emotion. Once fairness is understood in these terms, the near-impossibility of defining how to make others feel it is clearly apparent.

[483] ibid, 9.

[484] For an overview of possible regulatory models and their advantages and drawbacks see Nicolas Petit and Jerome De Cooman, 'Models of law and regulation for AI' in Anthony Elliott (ed), *The Routledge Social Science Handbook of AI* (London, Routledge, 2021) ch 13, 199; Cary Coglianese, 'Regulating Machine Learning: The Challenge of Heterogeneity' (2023) *Competition Policy International*, 9, ssrn.com/abstract=4368604.

[485] See Abigail Z Jacobs and Hanna Wallach, 'Measurement and fairness' (2021) *Proceedings of the 2021 ACM Conference an Fairness, Accountability, and Transparency* 375, 382:

> To date, much of the literature on fairness in computational systems has focused primarily on quantitative, parity-based operationalizations of fairness. These operationalizations are appealing to computer scientists because of their quantitative nature and because they operate within the boundaries of a single computational system, without reference to the broader societal context in which the system is situated. Yet these same properties mean that these operationalizations necessarily lack aspects of the substantive nature of fairness found in many of the theoretical understandings long discussed by philosophers, lawyers, and social scientists.

Second, human society looks for fairness in almost every aspect of human life, but at the same time does not want every aspect of life to be regulated. Thus the application of AI fairness regulation needs to be restricted. An obvious way to reduce the scope of application of regulation is to adopt a risk-based approach, applying regulation only to the most seriously risky activities. The difficulties here are two-fold. First, devising a risk taxonomy is fiendishly difficult, as the EU AI Act illustrates. As will be explained in section 8.6, using the hierarchy of risk set out therein would, so far as fairness is concerned, include some AI implementations which clearly pose few fairness risks while at the same time excluding from the regulation others which are seriously risky for fairness. Second, the degree of risk which an AI implementation poses is not derived primarily from its field of application, such as employment or medical diagnosis, but much more from its design assumptions and its development, training and testing processes. These are invisible to the external world, thus making the external assessment of risk near-impossible other than in the most general terms.

8.1.2.2. Co-regulation

Co-regulation is a commonly adopted regulatory model, developed most strongly for financial services regulation and often implemented for technology regulation.[486] Under that model, regulation imposes broad and largely undefined obligations on those who are regulated, the regulatees. The regulator responsible for that regulation develops detailed codes and guidance about how to comply with these obligations, identifying their content through 'conversational'[487] discussions with regulatees and external experts. These codes and guidelines are not strictly binding, but regulatees who comply with them have comfort that they do not run the risk of regulatory penalties if, in spite of their compliance, the regulator finds there has been a failure to achieve the broad obligations. Instead, the regulator requires the regulates to modify the technology and/or their systems of operation for the future, and updates the codes and guidance to ensure that others do not fail in similar ways. Failure to comply with codes, or to follow guidance without a clear justification, is likely to be judged a regulatory breach and thus attract regulatory sanctions. Thus in co-regulation the current means of compliance is always defined in some detail, but might change rapidly, requiring fresh compliance action by regulatees. Compliance is always a moving target. Clarke[488] advocates co-regulation as the primary regulatory model for AI. Here an AI regulator would be established in each jurisdiction, and that regulator would approve codes of conduct, developed in conjunction with the AI industry and other relevant

[486] Christopher T Marsden, *Internet co-regulation: European law, regulatory governance and legitimacy in cyberspace* (Cambridge, Cambridge University Press, 2011).
[487] Black (n 322).
[488] Roger Clarke, 'Regulatory alternatives for AI' (2019) 35 *Computer Law & Security Review* 398.

organisations such as ethics bodies. The regulator would have enforcement powers against AI producers who fail to comply with the code.

The theoretical advantages of co-regulation over the command and control model are obvious, in particular that the codes and guidance set out practicable ways to comply with the broad regulatory objectives. Uncertainty about the risks of adopting new technology is reduced because regulatees can be confident that enforcement action will not be taken if they follow the codes and guidance. Practically, though, co-regulation is only a useful model if the regulated domain is narrow enough for effective codes and guidance to be developed. It should be obvious from the preceding chapters that, for fairness at least, it is unlikely that a code for achieving compliance with a general fairness obligation could be developed at the high level of abstraction needed to cover all possible AI implementations. The fairness issues differ so radically between implementations that it is likely that hundreds of variations of codes would be needed, perhaps more, which if centralised in a single regulator would be likely to generate an impossibly large and unwieldy regulatory system. Such a wide range of potential compliance methods would be likely to generate contradictions and confusion in the minds of regulatees, and thus reduce the chance of AI fairness. These objections are less likely to apply to well-defined sectors of activity, and co-regulation is already the model adopted by most sectoral regulators. Many of these are already producing guidance and codes for the implementation of AI, as we saw in section 5.3.

Black and Murray[489] offer a further argument against the co-regulatory model. They argue that the current approach in most states, of relying on soft regulation in the form of ethical codes and recommendations, is inadequate. History shows that this approach failed for the internet, leading to a concentration of practical regulatory power in the hands of a small number of gatekeepers, the major technology companies such as Alphabet, Meta and Apple. The source of this concentration is the failure to identify regulatory goals at an early stage, and from those goals to produce regulation which takes account of the realities of the operations and incentives of those being regulated. Instead, gatekeepers were allowed to decide how the technology would develop, driven by their own internal aims and incentives, while regulation followed behind and attempted to deal with the individual harms which resulted. By adopting this approach, which is fundamental to evolving co-regulation, 'governments failed to appreciate that they had left markets to control a technology which relies on network effects and can create system wide risks and impacts.'[490]

In the case of AI fairness, I would argue that the clearest systemic risk is that the AI research community develops a consensus on how to achieve AI fairness which is unacceptably different from the human conception of fairness, and that

[489] Black and Murray (n 477).
[490] ibid, 6.

over time this consensus becomes so entrenched in AI development that it is not feasible for regulation to be devised which corrects the difference. As an example, we saw in chapter seven that current AI fairness techniques can be better than humans at detecting some kinds of bias, but that they are unable to cope with unexpected contextual information which a human would consider important to determine fairness. This means that some decisions might be fairer, and others less fair. The decision about whether this is an acceptable compromise in return for the benefits of AI is not one which should be made by the developer community, driven as it is by technical feasibility and the commercial demands of those who fund the AI's development.[491] The development of the internet gives us a clear example of how such an entrenchment might arise. In those early days it was technically most efficient for the initial internet technologies to rely entirely on self-identification by online actors. Inevitably, activity which creates societal problems moved online, such as criminal activity, anonymous online persecution, child access to gambling and pornography, and so on. Correcting those problems through law and regulation is proving challenging because the internet technologies cannot now easily be adapted to achieve reliable identification without widespread systemic disruption.

8.1.2.3. Regulatory Advance Approval

A regulatory model which has its origins in society's demand that a technology should be adequately safe is to establish a system for advance approval of that technology, which is required before it can be put into operation. States commonly use this model for technologies which present clear risks of injury to their citizens, such as motor vehicles and aircraft, and a useful example is found in the EU regulation of medical devices.

The Medical Devices Regulation[492] applies to any medical technology, including software, which falls within one of the classifications in the regulation. These classifications are based on the level of risk the device poses to a patient – thus wooden tongue depressors are in the lowest classification, whereas surgically invasive medical devices are classified much higher.

Before being used in medical treatment, the highest-risk devices must secure a conformity assessment from a conformity assessment body which has been notified under the relevant national law, to demonstrate that they meet the

[491] Black and Murray also note, ibid 7, that relying on ethical standards to shape technology development is unlikely to be a successful strategy:

> [E]thical standards for such systemic risks are insufficient, particularly in so far as they assume that risks are individualised and that the key to their management is the choices an individual consumer makes within the market place. Based upon our experience of internet regulation and governance from 1995 to today, such an approach will lead to future regulatory failures.

[492] Regulation (EU) 2017/745 of the European Parliament and of the Council of 5 April 2017 on medical devices, amending Directive 2001/83/EC, Regulation (EC) No 178/2002 and Regulation (EC) No 1223/2009 and repealing Council Directives 90/385/EEC and 93/42/EEC, OJ L117/1 5 May 2017.

requirements of the Regulations. The Regulations do not set out precise standards, because they would be different for every kind of medical device. Instead they set out in Annex I, paragraph 1 three fundamental principles which all medical devices must meet. They must:

- achieve the performance intended by their manufacturer;
- be designed and manufactured in such a way that, during normal conditions of use, they are suitable for their intended purpose;
- be safe and effective and not compromise the clinical condition or the safety of patients, or the safety and health of users or, where applicable, other persons, provided that any risks which may be associated with their use constitute acceptable risks when weighed against the benefits to the patient and are compatible with a high level of protection of health and safety, taking into account the generally acknowledged state of the art.

Additionally, both design and manufacture should take account of 'safety principles' (undefined) and the state of the art, as well as the environment in which the device is to be used and the capabilities and understanding of its intended users. 'Electronic programmable systems' (systems incorporating software, or software which qualifies on its own as a medical device) must be 'designed to ensure repeatability, reliability and performance in line with their intended use', and software must be 'developed and manufactured in accordance with the state of the art taking into account the principles of development life cycle, risk management, including information security, verification and validation.'

The entire focus of the regulatory regime is on risk balancing and risk management. Absolute safety and perfect performance are not required, as Annex 1, paragraph 2 makes clear: 'The requirement in this Schedule to reduce risks as far as possible means the reduction of risks as far as possible without adversely affecting the benefit-risk ratio.' In other words, medical devices are suitable to be put on to the market if they offer benefits to patients, and the risks which they create are not so great as to outweigh those benefits. This is inevitably a matter of judgement, but because a conformity assessment is required for each device the conformity assessment body which issues that assessment is making the initial judgement as an expert in the field.

The EU AI Act adopts this regulatory model for high-risk AI. AI which falls under existing product safety regimes, such as that for medical devices, needs to secure a conformity assessment from that sectoral regulator.[493] Other AI which falls within the Annex III classification of 'high risk' must, in addition to the other obligations placed on the AI 'provider' by the AI Act, secure a conformity assessment from a national approval body.[494] Annexes VI and VII set out the standards and processes for that assessment.

[493] EU AI Act (n 466), Art 6(1).
[494] ibid, Arts 16(e), 43.

As will be explained in section 8.6, the risk of unfair decisions does not map in any sensible way to the risk classifications in Annex III. The EU AI Act is clearly focused on physical health and safety and on fundamental human rights, and AI fairness seems to have received little attention. Even if fairness were specifically brought within the list of risks, the model of pre-approval via conformity assessment has much the same disadvantages as the co-regulatory model.

This is because the conformity assessment process is very much an implementation of co-regulation. The AI Act sets out broad standards, and the approval body is expected to develop detailed standards for particular AI implementations in consultation with the AI industry and external bodies. As is the case for co-regulation, the wide range of AI decision-making where fairness might be demanded will require an equivalently wide range of different standards. Contradiction and confusion seems inevitable.

Thus although the concept of requiring advance regulatory approval before implementing an AI to make decisions which affect third parties is an attractive one,[495] I reject it as a model for AI fairness regulation (outside existing regulated sectors, which might well find it useful) because its regulatory domain would be too wide for an effective approval system to be put in place.

8.1.2.4. Meta-regulation

There is one regulatory model which might be an appropriate foundation on which to build a system to regulate for AI fairness. This is meta-regulation,[496] which for our purposes is more precisely encapsulated by the idea of enforced self-regulation as conceptualised by Braithwaite[497] and by Gupta and Lad.[498]

In a meta-regulatory approach, the regulator does not regulate directly, but rather oversees the regulatory efforts of others and provides mechanisms to incentivise or enforce compliance with those alternative sources of regulation. Braithwaite describes this as 'the state acting as a facilitator or a monitor of corporate social control exercised by non-governmental institutions.'[499]

Meta-regulation focuses on an organisation's internal enforcement of its external regulatory obligations. Internal compliance enforcement by an organisation can be more effective than external review because those within the organisation have a shared corporate mission. Thus they exchange information more freely with

495 See Matthew U Scherer, 'Regulating Artificial Intelligence Systems: Risks, Challenges, Competencies, and Strategies' (2016) 29 *Harvard Journal of Law & Technology* 353.

496 Colin Scott, 'Regulating everything: From mega-to meta-regulation' (2012) 60 *Administration* 57; Peter Grabowsky, 'Meta-regulation' in Peter Drahos (ed), *Regulatory theory: foundations and applications* (Canberra, ANU Press, 2017), ch 9, 149.

497 John Braithwaite, 'Enforced self-regulation: a new strategy for corporate crime control' (1982) 80 *Michigan Law Review* 1466. Meta-regulation has strong conceptual similarities with what Coglianese describes as management-based regulation – Coglianese (n 484) 9.

498 Anil K Gupta and Lawrence J Lad, 'Industry self-regulation: an economic, organizational, and political analysis' (1983) 8 *Academic Management Review* 416.

499 Braithwaite (n 497) 1553.

each other, so that the internal compliance personnel have a deeper understanding and access to more information. However, a potential weakness of internal enforcement is that compliance recommendations can be ignored if they are costly or have long-term consequences. This is the reason why Braithwaite emphasises the need for external enforcement of those internal rules:

> Enforced self-regulation, on the other hand, can ensure that internal compliance groups will not be lightly overruled. Under the model proposed by this Article, a compliance director would be required to report to the relevant regulatory agency any management overruling of compliance group directives. A director who neglected this duty would be criminally liable. Such a provision would be the strongest method of ensuring that compliance unit recommendations would be followed by management. Companies that regularly ignored such directives would fall under the regulatory agency's special scrutiny. The agency could concentrate its limited prosecutorial resources on companies that continually and irresponsibly disregarded compliance group recommendations. Enforced self-regulation thus combines the versatility and flexibility of voluntary self-regulation, but avoids many of the inherent weaknesses of voluntarism.[500]

Braithwaite's proposal is aimed at corporate crime, and imposing criminal liability for non-compliance is clearly too extreme an approach for fairness regulation. However, a workable regulatory mechanism for fairness could be devised by taking three inspirations from his ideas:

- The broad regulatory requirement, in this case that the AI producer should use reasonable efforts to ensure that the AI does not make unfair decisions, should be translated into workable internal regulatory requirements by the regulatee itself.
- The regulatee should be required to report its internal regulatory efforts to the regulator and, as I will also suggest, to the wider public. This enables the quality of those efforts to be evaluated.
- Regulatory consequences should follow failure to establish appropriate internal regulatory standards, or to comply with them.

The advantages of meta-regulation with an enforcement element over the other regulatory models[501] include: flexible rules which are tailored to match the regulated entity; improved ways of achieving compliance; quick adjustment to changes in the business environment; fostering regulatory innovation; more comprehensive rule coverage; and that regulatees would be more committed to rules which they write themselves. The two main disadvantages which Braithwaite identifies in relation to corporate crime[502] are unlikely to arise when regulating fairness:

- Particularistic internal rules for fairness are unlikely to weaken the moral force of rules which should be universal, because we cannot identify a universal

[500] ibid, 1470.
[501] ibid, 1474–83.
[502] ibid, 1494–96.

fairness rule beyond 'no unfair decisions'. This rule has no detailed specification that might conflict with internal rules for achieving fairness.

- Detailed internal rules cannot be written to evade the spirit of the law because the regulatory system which will be proposed below focuses enforcement not on those rules, but on whether the internal rules amount to reasonable efforts to comply with the spirit of the broad fairness obligation.

Meta-regulation with enforcement is not a purely theoretical regulatory model. We can see it in action in some current regulatory systems which focus on technology. A particularly clear example is that way in which the US Federal Trade Commission (FTC) uses its powers[503] to enforce compliance by technology companies with their published privacy policies. In numerous cases the FTC has taken action against major players such as Google and Facebook, including imposing regulatory penalties, implementing ongoing audit and seeking injunctions breach of which is a contempt of court.[504] An important difference from Braithwaite's conception is that in the US there is no federal obligation to publish privacy policies. In the absence of a state law which does impose such a requirement,[505] the FTC only has power to enforce published policies, under the theory that it is an unfair practice for a business not to do that which it has promised to the public.

A second example which does impose a requirement to generate internal rules is the obligation under the EU GDPR, Article 35(1) for a data controller to conduct a data protection impact assessment where the planned processing poses 'a high risk to the rights and freedoms of natural persons'. Article 35(7) sets out what that assessment must encompass:

(a) a systematic description of the envisaged processing operations and the purposes of the processing, including, where applicable, the legitimate interest pursued by the controller;
(b) an assessment of the necessity and proportionality of the processing operations in relation to the purposes;
(c) an assessment of the risks to the rights and freedoms of data subjects referred to in paragraph 1; and
(d) the measures envisaged to address the risks, including safeguards, security measures and mechanisms to ensure the protection of personal data and to demonstrate compliance with this Regulation taking into account the rights and legitimate interests of data subjects and other persons concerned.

Failure to carry out an assessment where needed can lead to enforcement action, as can an inadequate assessment.

503 US Federal Trade Commission Act 1914, 15 USC § 45(a) (as amended), prohibiting 'unfair or deceptive acts or practices affecting commerce'.

504 For a brief overview see Asunción Esteve, 'The business of personal data: Google, Facebook, and privacy issues in the EU and the USA' (2017) 7 *International Data Privacy Law* 36, 38.

505 For example, the California Consumer Privacy Act 2018, California Civil Code s 1798.100.

Article 35 is, though, only one small meta-regulatory element of the GDPR. The majority of data protection regulation is still set out in the form of a command and control model. However, it should be clear that the concept of an impact assessment could be modified and used as the basis for a regulatory model for AI fairness.[506]

Most recently, initiatives to regulate the provision of online content in the EU and the UK have combined obligations to undertake risk assessments with obligations to take steps to mitigate those risks. The EU Digital Services Act 2022[507] imposes these obligations only on the largest online service providers,[508] in respect of 'any systemic risks in the Union stemming from the design or functioning of their service and its related systems, including algorithmic systems, or from the use made of their services.'[509]

Systemic risks are not defined,[510] thus leaving it to the platform to identify the full range of those risks, but Article 34(1) requires that the assessment must include the following risks:

- dissemination of illegal content;
- negative effects for the exercise of fundamental rights;
- negative effects on civil discourse, electoral processes and public security; and
- negative effects in relation to gender-based violence, the protection of public health and minors and serious negative consequences to the person's physical and mental well-being.

Once systemic risks have been identified, Article 35 requires providers to put in place 'reasonable, proportionate and effective mitigation measures, tailored to the specific systemic risks identified.'

The focus is clearly on online content which is potentially harmful to individuals and society, and the regulatory aim is that these service providers should develop their own systems for identifying, monitoring and mitigating the risks created by such content.

[506] Indeed, the concept of such a model has already been proposed in Lilian Edwards and Michael Veale, 'Enslaving the algorithm: From a 'right to an explanation to a 'right to better decisions?' (2018) 16 *IEEE Security & Privacy* 46. See also Jamy Li and Mark Chignell, 'FMEA-AI: AI fairness impact assessment using failure mode and effects analysis' (2022) *AI and Ethics* 1, adopting additional ideas from failure analysis in engineering and manufacturing design and production.

[507] Regulation (EU) 2022/2065 on a Single Market For Digital Services and amending Directive 2000/31/EC (Digital Services Act), OJ L277/1 19 October 2022.

[508] 'Very large online platforms', which are those designated by the European Commission under Art 33(4) on the basis of their average monthly active service recipients, currently set at 45 million by Art 33(1).

[509] Digital Services Act 2022, Art 34(1).

[510] Recital 79 suggests that the risk assessment should focus on societal and economic harm, but does not attempt to narrow that down to explain where the line between systemic and non-systemic risks should be drawn.

Compliance is in part assessed through annual, independent audits of compliance under Article 37, and following each audit the results of the risk assessment and the mitigations adopted must be published.[511] National regulators (Digital Services Coordinators) also oversee compliance and have powers to accept undertakings, order cessation of infringements and impose remedies, or issue fines/penalties or bring prosecutions.[512]

The UK's Online Safety Act 2023 is more ambitious, applying to all but the smallest online service providers. In addition to imposing obligations to prevent the most egregious forms of online content[513] from being disseminated via their services, there are risk assessment and mitigation obligations which differ between adult users and children.

Risk assessment and mitigation in respect of illegal content applies for all users.[514] For children, however, the obligation is much broader, extending to any content which is harmful to children,[515] and the risk of harm has to be assessed in relation to their age and any particular vulnerabilities they have. The UK regulator, OFCOM, has powers under section 99 to issue guidance and an obligation under section 41 to develop codes of practice, and has already done so in draft for consultation in respect of illegal content.[516] OFCOM also has extensive enforcement powers.[517]

Although at first sight these aspects of the Online Safety Act look very like a meta-regulatory system, the long-term intent is clearly to use the reports of risk assessments and mitigation measures to develop a co-regulatory system for online safety. This is particularly clear from section 49, which provides that a service provider which adopts a means of compliance recommend in a code of practice is to be treated as complying with its duty under the Act. Once codes of practice have been finalised, it will be a brave service provider that decides to adopt some alternative, at least without prior discussion with and reassurance from OFCOM.

[511] ibid, Art 42(4).

[512] ibid, Art 51(2).

[513] For all users this consists of 'priority' illegal content, defined in s 59(1) as content relating to terrorism, child sexual exploitation and abuse, or an offence specified in Sch 7. Where online services are likely to be accessed by children this extends to 'primary priority content that is harmful to children', defined in s 61 as:

- pornographic content;
- content which encourages, promotes or provides instructions for suicide;
- content which encourages, promotes or provides instructions for an act of deliberate self-injury;
- content which encourages, promotes or provides instructions for an eating disorder or behaviours associated with an eating disorder.

[514] Online Safety Act 2023, ss 9, 10, 26, 27.

[515] ibid, ss 11–12, 28–29. Content harmful to children is defined in s 60. Harm is a very broad concept, covering both physical and psychological harm – s 234(2).

[516] www.ofcom.org.uk/consultations-and-statements/category-1/protecting-people-from-illegal-content-online.

[517] ibid ss 130–151.

8.1.3. Regulatory Obligations

The purpose of regulation, as Black puts it, is 'attempting to control, order or influence the behaviour of others'.[518] Thus we need to decide what AI fairness obligations we might usefully impose by regulation in an attempt to achieve such control, order or influence.

To start with, we should dispose of the apparently easy solution. This is simply to require that some – maybe even all – AIs should act fairly, without seeking to explain in any further detail what that demand requires. This easy answer is tempting because it is how we currently regulate humans when we require them to act fairly (see section 5.2), and also how we deal with other hard-to-define social demands such as reasonableness[519] and safety.[520] This approach places the regulatory obligation to achieve fairness on a person, most likely the decision-maker, foreseer of risk of unfairness or AI producer, and assumes that the person subject to the obligation:

- understands what outcome needs to be achieved, because that outcome (fairness in our case) is a shared human conception of the concept; and
- is capable of achieving that outcome (or if not, should not engage in the activity). A person who cannot drive with reasonable care should not drive at all, and a product producer should not market unsafe products.

Neither of these assumptions holds for AI decision-making. First, although the human producers of the AI might well have an adequate emotional understanding of how to make fair decisions, AI does not make decisions in the same way that humans do. We saw in chapter seven the difficulties which AI developers face in translating the human, emotional conception of fairness into techniques which might approximate the same decision outcomes. Even if they adopt the simplified conceptions of that understanding which philosophers have analysed, AI developers are still left with numerous choices, each of which leads to different fairness outcomes. At best, there is real uncertainty how AI fairness ought to be achieved, and it is almost certain that any AI decision system will produce at least some unfair decisions.

Second, there may be good reasons why society should accept some level of failure to achieve AI fairness, particularly if mechanisms can be put in place to remedy the worst failures. There are clear benefits to be achieved from using AI in many cases, but regulation which potentially imposes liabilities which cannot be mitigated through good practice will deter AI producers and thus prevent those

[518] Julia Black, 'Critical Reflections on Regulation' (2002) 27 *Australian Journal of Legal Philosophy* 1, 25.

[519] Via, for example, the tort of negligence.

[520] See eg EU Product Liability Directive 1985 85/374/EEC, Art 6(1): 'A product is defective when it does not provide the safety which a person is entitled to expect, taking all circumstances into account …'.

benefits from being achieved. We recognise this for human activities – products must be safe enough, not absolutely safe, and drivers need only be reasonably careful. Failures will occur, but society accepts that some level of failure must be allowed for.

Another solution, which we can also dispose of quite briefly, is that we should specify in detail what AI producers are required to do, imposing some form of liability or sanction if they do not. The previous chapters have demonstrated at length how difficult it is to break down the human conception of fairness into individual components. All humans might feel the emotion of fairness, but in every specific case there is potential for disagreement about whether a decision engenders that emotion. Each of us puts a different value on particular instances of, for example, merit and disadvantage, or power and status, and our ultimate judgement about fairness comes from the influence each plays on our emotions. That judgement is also swayed by the context in which the decision is made, and context is almost infinitely variable.

A further obstacle to specifying fairness obligations in detail is that the range of AI applications is vast. It seems to me certain that each reader of this book would judge the fairness of an AI loan application system in quite a different way from the fairness of an AI which allocates scarce medical treatment. For example, the effect on the future life of the patient of a positive decision would clearly be a relevant fairness matter for the medical treatment system, but we would be unlikely to consider the future life of the applicant when assessing whether a loan application was decided fairly.

This tells us that we would either have to specify the fairness obligations of AI producers in such general terms that the regulation, in effect, said almost nothing useful, or in such specific detail that we would be trying to predict every application and every likely decision context, which would be impossible. Neither is likely to achieve any socially useful purpose. And because effective regulation demands that its regulatory standards should be both understandable and achievable, neither approach is likely to be effective in achieving fair AI decisions.

The approach to regulation which I will propose is therefore a meta-regulatory approach which, rather than defining the content of the fairness obligation, instead requires the AI producer to identify the fairness needs of their AI implementation and then to ensure that the AI is built to and operates in accordance with those needs. This is not dissimilar in overall concept to the suggestions made by Edwards and Veale.[521] They review the right to a retrospective explanation for automated decision-making under Article 22 of the EU GDPR, together with proposals to improve that right, and conclude that those explanations of AI decision-making are not as helpful as expected in achieving AI fairness and lack of bias. The onus is placed on users to obtain an explanation, and once users are given that explanation it is hard to understand. Even if understood, it may not identify

[521] Edwards and Veale (n 506).

fairness defects in the AI which are only discoverable from a deeper analysis of the system as whole. Finally, and most importantly, the right to an explanation provides little incentive to AI producers to ensure fair decisions in the first place if only a minority are ever likely to be challenged (particularly if only a small proportion of that minority are challenged successfully because of the difficulties in identifying regulatory failure from retrospective explanations). Instead, they propose a new right to operate at the design or deployment stage, aimed at achieving better (ie fairer and less biased) decisions. They suggest this can be achieved by requiring impact statements, based on the concept of a data protection impact assessment, combined with third-party certification. In effect, the AI producer would be obliged to identify fairness risks and modify their design to mitigate them, and third-party certification would provide an assurance that this has been done at a satisfactory level.

I do not include third-party certification because it is a form of advance approval, and therefore has the defects discussed in section 8.1.2.3. I do, though, focus on enforcement, because I recognise that any effective technology regulation system needs to provide differential levels of enforcement for different types of risk. As Black and Murray point out:

> In general, where risks fall on individuals and can be compensated for, then regulation is comprised of ex post liability regimes (e.g. negligence, contract, statutory regimes of product liability or food safety) which may or may not be supplemented by state or non-state based oversight and enforcement. This is the 'develop, deploy, regulate' model noted above. Where the risks of individual harm are such that it is considered a more precautionary approach is required, then those making such products or delivering such services may be required to have specific authorisation (financial services, for example) (develop, regulate, deploy). Where the risks fall on individuals but are non-compensatable or 'deep regret', such as threat to life, then there more onerous ex ante requirements are imposed through licensing and ongoing monitoring and enforcement, and standards of consent to be exposed to the risk are higher (such as licensing of pharmaceuticals and consent to medical treatment). On the other hand, where risks are systemic even though still compensatable, then systems which rely on individual consent to the risk are inadequate, and again ex ante regulation is required as well as ex post remediation. Payment systems are a good example, and indeed the argument above is that the internet should have been seen as posing such systemic risks. Finally, where the risks are systemic and the harms non-compensatable or non-remediatable, then regulatory regimes are ex ante and usually highly restrictive, with regulation of development imposed as well as deployment, requiring extensive trialling and close regulation: such as the use of genetically modified organisms, stem cell treatments, or in the case of aviation or nuclear power.[522]

The immediate consequences of an unfair AI decision will be felt by the decision subject, which suggests that compensation should be the primary consequence of regulatory breach. However, in many areas of activity an unfair

[522] Black and Murray (n 477) 13.

decision system poses systemic risks, particularly to the societal acceptance of that activity – financial services is an obvious example – and so here additional regulatory enforcement sanctions may be needed.[523] The risks of unfairness are not usually non-compensatable or non-remediable, so restrictive regulation pre-deployment would seem excessive.

8.2. Aims of the Regulatory Model for Fairness

Any regulatory model for AI fairness needs to answer three questions:

- Who should have the responsibility to achieve fairness?
- How can we frame fairness regulation so that it has some chance to achieve its aims? An important issue here is whether the problem, and its solution, is clear enough that a definitive answer can be given. If not, the question becomes whether it is possible to identify some initial solution which is capable of future development as clarity about problems and solutions emerges.
- What should the scope of regulation be? Should it apply to every AI implementation or only some?

In order to answer these questions I suggest that we should stand back for a moment from considering the model and content of regulation and ask ourselves instead how we currently regulate human decisions for fairness. The legal approach to fairness obligations was analysed in section 5.1, and we can break it down as follows:

- The aim of fairness is to treat like cases in a like manner, potentially making some allowance for merit and disadvantage. How to account for merit and disadvantage is left largely to the decision-maker's discretion, recognising that there is no consensus on this question.
- The human decision-maker needs to aspire to act fairly. This aspiration is demonstrated by adopting relevant merit criteria, identifying the factors which are relevant to the decision, gathering accurate information about them, excluding irrelevant factors and information from the decision, and weighing all these together in good faith to come to a conclusion. It is worth noting that legal challenges on grounds of fairness usually only reconsider the conclusion if it is clear that there was no good faith attempt to achieve fairness, or that the decision-maker failed to follow its own established decision-making procedures. This recognises that the human conception does not establish a fixed and certain standard. Challenges succeed on the basis that the decision-maker

[523] This possibility also enhances the likelihood of compliance, because reliance on compensation alone has been shown to be less effective in achieving that end – see Robert Baldwin, Martin Lodge and Martin Cave, *Understanding Regulation* (Oxford, Oxford University Press, 2013).

has not demonstrated their aspiration to decide fairly, for example because they have considered irrelevant material or failed to gather accurate information.

- The human decision-maker must go about the decision-making in a way which is likely to achieve the fairness aspiration, and which is likely to convince observers on that point. This encompasses procedural elements such as offering a fair hearing or otherwise giving the subject the chance to provide relevant information, and excluding from decision-making those who might be biased or risk the appearance of potential bias.

What this boils down to is an external assessment of whether the decision-maker has acted in good faith and made reasonable efforts in their attempt to achieve the overall aim of fairness. Could we base a new system of AI fairness regulation on this approach?

Obviously we cannot ask these questions about the AI system itself. As a technology, AI has no conception of good faith, and asking about its efforts produces a meaningless answer. However, we *can* ask these questions about the humans who are ultimately responsible for the AI's decisions – the AI developer and the AI user, collectively the AI producer. If those humans have acted in good faith and made reasonable efforts to ensure their AI makes fair decisions, this offers a reasonably high level of assurance to society that those decisions are in large part likely to be as fair as is achievable[524] even if perfection is not attained. If those humans explain what they have done, and their reasoning for doing so, external observers can assess the quality of their efforts and, if necessary, request or demand improvement and even ask for redress.

In what follows I will put forward an outline for a regulatory system based on this approach. The regulatory standards are those which already apply to human decision-makers, and so should be understandable and achievable by AI producers. I will also explain how the system is both flexible and expandable, so that it can provide a basis for the more detailed regulation of fairness which we might desire in particular sectors of activity.

8.3. Who to Regulate?

In common parlance we speak of 'regulating AI', but if taken literally that is nonsensical. AI, like all technologies, is completely insensitive to regulation. It cannot hear the commands of the regulator about how it should carry out its activities, nor can it change its 'behaviour' in response if it did hear those commands.

[524] The qualitative investigation into attitudes to algorithmic fairness by Woodruff et al (n 86) 7 found that minor deviations from statistical fairness were considered to be acceptable; 'participants appeared to interpret small statistical disparities as benign, largely considering them to be natural, inevitable, and impossible to fix.'

Thus regulation needs to be addressed to some person[525] who has the power to ensure that the AI operates in the way desired by regulation, making changes if necessary.[526]

There are three factors to consider when deciding which person or persons should be subject to fairness regulation:

- Who does society at large consider to be responsible for the AI's decision-making activities? Our social conception of justice considers that responsibility justifies the imposition of duties by society, and conversely that it is wrong to impose duties on those who are not responsible.
- Who is the least cost avoider? In other words, who is the person who will find it easiest and least burdensome to ensure the regulation is complied with? Placing the regulatory burden on that person is the most economically efficient solution, and the lower the cost (in time, effort and money) the more likely that compliance will be achieved.
- If there is a choice of responsible persons, are there pragmatic considerations to guide that choice? The most obvious of these is the ability to enforce the regulation, which is in particular affected by the location of the relevant persons. If one such person is located in the jurisdiction of regulation and the other outside, it is obviously easier to enforce locally than abroad. It is also worth noting that extraterritorial regulation has far less authority than regulation which applies purely within the jurisdiction, and the strength of its authority is directly connected to the likelihood of voluntary compliance.[527]

Throughout this book I have focused on the AI developer and the AI user, and these persons clearly share responsibility for the AI's decision-making. The question is therefore whether fairness regulation should apply to one or the other, or both.[528]

The answer to that question depends in part on how responsibility is shared between them. This will vary in different implementation scenarios. It will be

[525] I initially wrote 'human', but law and regulation is often addressed to corporations, who are non-human legal persons. However, corporations can only act through their human officers, employees and agents, so ultimately it will be a human who has to effect change to the AI.

[526] This is true even for generative AI. It is conceivable that a lawmaker might address such an AI directly, for example instructing a generative AI like ChatGPT that in all its future operations it should not refer to particular types of content in its responses to users. ChatGPT might conceivably respond to the lawmaker that it agrees to do so, but it would not actually modify its behaviour as a result. OpenAI, the developer, could modify its operations to comply, or perhaps change the learning process to give more weight to the instructions of lawmakers, but ChatGPT could not do so autonomously.

[527] Murray and I have examined this issue at length in relation to the cyberspace technologies – see Reed and Murray (n 23).

[528] In what follows, I write of AI developer and user as if they are clearly separate entities. It is not uncommon, though, for AI development to be conducted entirely in house by an organisation. However, there will still be a distinct functional and responsibility split between those who design, train and test the AI (developer team), and those who determine when and how it will be implemented and used by the organisation (user team).

helpful to run through three likely scenarios, taking as our example a recruitment selection AI.

One possibility is that the AI developer produces a general purpose recruitment selection AI and makes it available via an AIaaS model. Here, the developer has sole responsibility for the design, training and testing of the AI, and determines the fairness techniques which will be incorporated into it. The AI user is responsible for selecting this particular AI to use for its recruitment selection, from those available from other AIaaS providers, and also for determining the merit criteria against which the AI will judge candidates. The user is also responsible for collecting and inputting the data about candidates which drive the AI's decision.

A second possibility is that the developer offers some degree of customisation to users. In particular, there might be a range of fairness techniques which could be incorporated into the AI from which the user can select. The user might provide training data, derived from its previous recruitment activities, but the developer will be responsible for checking the adequacy of that data and for conducting the training and testing processes. Merit criteria and the collection and input of candidate data will remain the responsibility of the AI user. There is a high level of shared responsibility here.

Our third possibility is a bespoke AI development. Here the developer and user will share responsibility for almost all the matters discussed above, probably including the setting of merit criteria, though the AI user will be solely responsible for the collection and input of candidate data for individual decisions once the AI is put into service.

We can see that there is no clear dividing line of responsibility between AI developer and AI user,[529] which therefore suggests that we might want to apply our fairness regulation to both. It is also unclear which of them would be the least cost avoider in terms of complying with regulation. The AI developer will ultimately need to implement any regulatory fairness requirements, but the AI user is paying the developer as its customer and thus has the practical ability to require the developer's assistance in achieving regulatory compliance.

The pragmatic questions of enforceability and regulatory authority may thus be the determining factors. Sensible lawmakers try to avoid making laws and regulations which have substantial extraterritorial effects for precisely these reasons. In all three of our examples the AI developer might be located anywhere in the world, whilst the AI user is more likely to be located in the regulatory jurisdiction. Additionally, the sensible lawmaker will try to limit the application of regulation so that it applies only to AI decisions which affects the rights and interests of decision subjects in the jurisdiction. The AI developer may have

[529] For a deeper analysis of how responsibility is shared across the AI supply chain see Jennifer Cobbe, Michael Veale and Jatinder Singh, 'Understanding accountability in algorithmic supply chains' (2023) *Proceedings of the 2023 ACM Conference on Fairness, Accountability, and Transparency* 1186.

no idea where potential decision subjects are likely to be located, whilst the AI user will always know their location (or at least, in our example, the location of the place of potential employment, which seems the place where rights and interests should be considered).

Thus the pragmatic answer might be to apply the regulation primarily to the AI user, leaving the user to persuade the AI developer to provide appropriate assistance, particularly via the contract terms between them. This would place all the compensatory and regulatory liability risks on the AI user, an approach which mirrors existing frameworks of legal responsibility. Those who use products or services produced or provided by another in activities which affect third parties are not thereby absolved from responsibility. They already have obligations to take care to ensure that the product or service was appropriate for that use, and to use it responsibly. Thus an organisation or individual which implements an AI system developed by another should be expected as a matter of good practice to assess whether the AI is suitable for the task, and to ensure that the way in which it is deployed does not impose unreasonable risks on others. A regulatory fairness obligation which requires what should already have been done can hardly be objected to.

A practical objection to this approach is that there is often a substantial informational and power imbalance in favour of the AI developer, and so the AI user may be unable to negotiate appropriate contract terms which require the developer to provide the information about fairness which the user needs, or to agree a suitable risk allocation for compensatory liability. The simplest way to mitigate this problem is by extending any existing unfair contract terms controls so that they apply to such contracts, irrespective of any choice of applicable law. The UK approach under the Unfair Contract Terms Act 1977 would be to subject terms which exclude or limit the AI developer's liability to a test of reasonableness, and the terms in question would be the limitation or exclusion of liability for (a) failure to provide an appropriate risk mitigation explanation (see section 8.4.1), or (b) failure of the AI to perform in accordance with that explanation. This would offer flexibility to accommodate the broad range of AI developments. For example, it would be reasonable for the developer of a foundation model generative AI, like ChatGPT, to provide a generic explanation of the fairness risks which were identified and mitigated in the development, but no explanation about how the AI deals with any fairness issues which arise purely from the AI user's intended implementation. Following the same reasoning, a standard cap on liability for non-conforming performance of a foundation model AI might also be reasonable. By contrast, in a bespoke AI development for the user it is reasonable to expect a more focused risk mitigation explanation from the developer, and probably a higher level of liability acceptance.

However, there may be circumstances in which the fairness risks which the AI presents are so great to society, such as an AI which makes criminal justice decisions, that at least some elements of the regulation should apply to the AI developer even if that would result in extraterritorial application of the regulation.

This can only be determined as a policy decision by the state in question, taking into account factors such as the level of risk, the need to achieve social acceptance, and the effect of the regulation on relations with other states.

In most cases, therefore, the regulatory obligation should be placed on the AI user. That obligation should extend also to the AI developer when the policy of the state requires it, based on the level of risk which that use of AI creates. For the remainder of this chapter I will revert to using 'AI producer' which encompasses both, but readers should remember that in most cases the user will be the primary subject of the regulation.

8.4. The Obligation to Explain and Mitigate Risks of Unfairness

Once we have decided which person is responsible under the regulation, the next question is how we should specify regulatory obligations which, if complied with, will demonstrate that person has acted in good faith and made reasonable efforts in their attempt to ensure that their AI will achieve fair decision-making. We have already seen that that it is not practicable to try to specify how fairness should be achieved through detailed rules, both because the AI fairness techniques available achieve different conceptions of fairness and because different types of decision system present different fairness issues. We have also seen that our current legal and regulatory approach concentrates not on achieving fairness, but rather on avoiding unfairness. Thus the regulatory obligation should be based on a duty to identify the risks to fairness which the AI decision system creates, to take reasonable steps to mitigate them, and to explain and justify how those steps are likely to mitigate the risks of unfair decision-making.

I suggest this is best achieved at two levels: a general obligation to produce a 'risk mitigation explanation' focused on fairness,[530] and more detailed obligations at a sectoral level. The risk mitigation explanation is the producer's analysis of the risks generated by the AI, together with a justification for the steps taken to reduce those risks.

8.4.1. The Basic Obligation

There are three reasons why requiring a risk mitigation explanation from AI producers, where the risks generated by their AI are sufficiently serious, is likely

[530] This is conceptually very similar to Leslie's suggestion of a Fairness Position Statement; see Leslie (n 103). I expand that idea to include a justification for the fairness choices which have been made and an obligation to ensure that the AI operates in accordance with it, and set out potential consequences for failure to do so.

to achieve the regulatory aim of fairer AI decision-making. First, it obliges the producer to explicitly state not only what has been done to avoid unfair decision-making but also the reasons and the justifications for so doing. Blindly following 'industry best practice' or some external guidance, even if such things exist, is clearly inadequate when it comes to explaining and justifying. The mere act of constructing such an explanation makes it more likely that the AI producer will make a genuine attempt to mitigate the risks of unfairness, if only because it forces them to identify and confront those risks. Additionally, as we saw in section 4.1.6, a person who explains in advance how they propose to achieve fairness also imposes on themselves a normative, internal obligation to do what they said they would do.[531] The justification of the risk mitigation measures taken is a core element of this approach to regulation because it forces the AI producer to construct an argument as to why society should accept their AI as being adequately fair. That argument will need to be defensible because of the potential consequences of producing an inadequate risk mitigation explanation (see section 8.7), and thus producing it is far from a conceptual or paper exercise.[532]

Second, because the risk mitigation explanation sets out the fairness aims of the AI producer and justifies how they will be achieved, this provides a standard against which external observers can assess how far the producer has in fact acted in good faith and made a reasonable attempt. Decision subjects and wider society will use that explanation to decide whether it generates in them the emotion of fairness, which is what is required for social acceptance of that use of AI. Regulators will be able to evaluate whether the producer claims to have approached the fairness problem in an appropriate way without needing to investigate the internal workings of the AI technology and its wider decision system. The evidence which an AI producer puts forward to support its justifications will play an important role here.[533]

Third, the explanation also provides a standard against which to assess whether the AI producer has done what they promised in the explanation. If the AI does not conform to the explanation in the way it makes decisions, that clearly

[531] See Peters (n 167) and accompanying text.

[532] A similar suggestion was published by Burr and Leslie as this book as being finalised, adopting the term 'argument-based assurance', which is:

> a process of using structured argumentation to provide assurance to another party (or parties) that a particular claim (or set of related claims) about a property of a system is warranted given the available evidence.

Christopher Burr and David Leslie, 'Ethical assurance: a practical approach to the responsible design, development, and deployment of data-driven technologies' (2023) 3 *AI and Ethics* 73, 74. Defensibility of the argument is a key element here, requiring evidence which supports the argument and allows it to be evaluated by external stakeholders (ibid, 91–93).

[533] For a comprehensive survey of mechanisms which an AI producer might use to provide evidence see Miles Brundage et al, 'Toward trustworthy AI development: mechanisms for supporting verifiable claims' (2020) arXiv preprint arXiv:2004.07213.

demonstrates a failure on the part of the producer to ensure the AI does what has been promised. If a liability claim is made by some person who alleges they have been treated unfairly, the explanation will set out the standards against which performance of the AI can be assessed.

The previous chapters have identified the issues which an AI producer will need to deal with in its risk mitigation explanation, and these can be summarised as follows. The producer will need to:

- identify and explain the risks created by their AI to the fundamental rights and economic interests of decision subjects;
- explain the measures taken to mitigate those risks, which would include, where a choice of risk mitigation measures is available such as in bias detection and correction, making explicit which choices were made;
- explain the measures taken to ensure that training and testing of the AI are based on data which are sufficiently accurate and complete, and to ensure that the input data when the AI is operated are also sufficiently accurate and complete.
- justify the AI producer's expectation that the measures taken will mitigate the risk appropriately.
- take reasonable steps to ensure that the AI operates in accordance with that explanation.

Requiring AI producers to construct such explanations does not seem an unreasonable burden to impose on them. AI developers are already exhorted to incorporate ethical compliance, such as fairness, into the design process, and methodologies and toolkits to assist them have been developed.[534] AI users can obtain this information from the AI developer, and then assess their own areas of responsibility (such as setting merit criteria and collecting and inputting data about decision subjects) to identify risks and mitigations.

The risk mitigation explanation will need to be made publicly available, both to increase social acceptance of that AI implementation and to ensure that those who are affected by the AI's actions can discover the standards which they are entitled to expect from it. Public availability also ensures that the explanation is devised before the AI is used to make decisions affecting individuals, rather than rapidly constructed after the event if, say, there is a regulatory demand or a threat of litigation.

To ensure that AI producers construct these explanations there will, of course, need to be consequences for failure. These are discussed in section 8.7.

Finally, we should note that AI producers are likely to need guidance about how to construct their risk mitigation explanations. Good faith and reasonable

[534] See eg Batya Friedman, 'Value-sensitive design' (1996) 3 *interactions* 16; Jiehuang Zhang and Han Yu, 'EID: Facilitating explainable AI design discussions in team-based settings' (2023) 7 *International Journal of Crowd Science* 47.

efforts clearly requires them to take account of ethical guidelines such as those discussed in section 4.1.1. However, more detailed guidance is likely to be helpful – this might be set out in the regulation itself, be provided by external authorities such as industry bodies or international organisations, or of course produced by any regulator which is appointed to oversee AI fairness. At the level of the regulated sectors we can expect to see more detailed and more focused guidance.

In sectoral guidance in particular we might expect that guidance to provide some assistance to the AI producer in deciding how far to concentrate on individual, as opposed to group, fairness. We saw in section 7.5 that there are arguments that a group focus can produce a socially acceptable level of individual fairness, particularly if a wide range of context is taken into account. The human conception of fairness requires individual fairness, whilst most AI development techniques for achieving fairness are inevitably based on some level of abstraction into groups. The regulator is better placed to identify the socially acceptable level of fairness than are individual AI developers and users.

8.4.2. Sectoral Specifics

Every jurisdiction has decided to impose specific regulation on some sectors of activity, and does so because that activity creates risks to society and individuals which are thought to be best controlled through regulation. Broadly speaking, some regulators focus on risks to wider society (such as transport and environmental safety), others regulate mainly risks to individuals (eg data protection and privacy), and many are charged with risk management of both kinds. Those regulators whose remit includes the risk of unfairness to individuals will most likely need to add specific detail which expands the scope of the general obligation to produce a risk mitigation explanation for AI decision systems. There are three areas in which such expansion might be expected.

The first is to specify in more detail the fairness risks to individuals' fundamental rights and economic interests which AI producers ought to foresee and guard against. As an example, in section 5.3.3 we examined the regulation and guidance issued by the FCA, which aims to ensure that customers of financial services firms are treated fairly, and these specify a number of consumer outcomes which those firms should aim to achieve.[535] Those outcomes clearly identify a number of fairness risks to consumers:

- that they are sold financial products which do not meet their needs;
- that they are not properly informed and do not understand the product;
- that they receive poor advice from the firm;

[535] n 318.

- that financial products and services do not perform as expected; and
- that there are barriers to consumer complaints or claims, or to switching to other providers.

These risks are generic to all financial services products, and so an AI producer will need to consider them when developing or implementing the AI and guard against their arising. As AI becomes more widely implemented in financial services the sectoral regulator is likely to identify more specific risks which AI generates for each financial product. Specifying that AI producers must take them into account will enhance the likelihood that those AI systems produce fair outcomes.

Guidance is likely to be useful to explain how these risks should be managed for particular financial products, because the nature of the risk is different for each. If we imagine that a financial services firm develops an AI to guide consumers through its range of products, the fairness risk that the customer is not properly informed and does not understand the product needs to be managed differently for, say, unsecured loans as opposed to spread betting. A consumer taking out an unsecured loan needs to understand little more than the consequences of failure to repay and the interest rate. By contrast, spread betting can generate substantial losses very quickly, something which consumers find hard to understand. It would likely be unfair for an AI which failed to explain to and warn consumers of risks they might not expect, and to assess their risk appetite appropriately, to recommend such a product to a consumer.[536] Guidelines from the regulator on how AI might assess risk appetite would clearly be of value here.

A second area in which sectoral regulators might want to expand the scope of the general regulatory obligation is in relation to the context of decisions. We saw in the preceding chapters that the human conception of fairness requires a decision-maker to consider the complete relevant context before making a decision, and this should include contextual elements which have not previously been identified if external observers would think that not considering them would be unfair.[537] However, we also saw in section 7.6 that the development process for AI needs to identify relevant context in advance, so that it can be labelled in the data and used in the AI training process. Thus an AI system will be unable to give weight to context which has not been foreseen.

By specifying contextual factors which AI producers ought to incorporate, sectoral regulators can at least improve the fairness of AI decision outcomes by ensuring that AI producers include data on those factors in their training, and

[536] See UK Financial Services Ombudsman, *Spread betting and contracts for difference*, www.financial-ombudsman.org.uk/businesses/complaints-deal/investments/spread-betting-contracts-difference.

[537] Wachter et al (n 473) 28 describe this contextual flexibility as a deliberate design decision about law.

ensuring that AI users collect data about them as input to actual decisions. Regulators can also specify contextual factors which ought not to be considered, because of the known risk that doing so might result in unfairness.[538] The most important contextual factors will become apparent over time, as complaints about unfair decisions are made, and can be incorporated into specific regulation as factors which must be incorporated, or more flexibly in guidelines which can be changed more rapidly and which would suggest factors which AI producers ought at least to consider incorporating. If we think about the problem of an AI assessing a consumer's risk appetite before recommending a spread betting product, some relevant elements of context are identifiable now. These include the consumer's net wealth, their current financial commitments and their investment history. Future adjudications on complaints might identify (purely hypothetically) that the consumer's long-term health prospects are relevant,[539] and if so the regulation or guidance could be updated accordingly.

Third, regulators commonly give guidance about their expectations to regulatees. This guidance will obviously differ between sectors, and between different activities within each sector. We reviewed above how the FCA guidelines on the fairness obligation explain that the regulator expects to see a range of fairness outcomes. The ICO guidance (see section 5.3.2.2) sets out different expectations, focusing on individual rather than statistical accuracy, avoiding unlawful discrimination and meeting a data subject's reasonable expectations. This kind of guidance is clearly helpful to AI producers – if their fairness mitigation explanations satisfy the regulator's expectations, fairness is more likely to be achieved and the risk of regulatory consequences is reduced or avoided.

8.5. Review and Appeal Processes

Judicial processes normally build in mechanisms to correct defects in the original decision-making via rights to ask for a review by, or appeal to, an independent person or body. Thus an obvious way to enhance AI fairness is by incorporating analogous mechanisms.

At least in the present state of the art, that appeal or review would be best conducted by a human. This is because of the emotional nature of fairness (or more accurately unfairness, which is what the reviewer would be focusing on). In particular, many fairness appeals would be arguing that the AI ought to have taken into account contextual factors which were not considered because they were not

[538] Chockler and Halpern (n 442) analyse in some depth potential factors which a financial regulator might consider for inclusion and exclusion.

[539] For example, a consumer with a history of cautious investment, but who has recently been diagnosed with a terminal illness and has no dependants, might reasonably decide to take high risks with wealth they have no prospect of consuming.

designed into the AI's decision processes. At the moment, only human reviewers have sufficient flexibility to evaluate arguments of this kind.

However, it is clearly inappropriate to require some form of fairness appeal or review from every AI decision. The reasons for implementing AI decision-making include speed, the ability to make more decisions, and cost. Human review removes these advantages. Thus we might expect to see human review processes only where:

- some regulator considers that the potential harm to decision subjects arising from unfair AI decisions is so serious that a correcting mechanism is needed; and
- in those circumstances where the AI user itself sees advantages in offering review or appeal. An example here might be where an organisation implements an AI to identify which of its staff should be considered for promotion. Some rejected candidates might well feel aggrieved by what they would consider the purely mechanical evaluation of their abilities, and might wish to put forward additional skills and achievements which the AI appeared to ignore. A dissatisfied workforce which does not feel it has been treated fairly is less likely to do its work well, and staff turnover might increase. The time and effort required for human review might well be worthwhile for the organisation if this is likely.

A human review or appeal might be based on two grounds. The first would be some inadequacy in the fairness mitigation explanation, if one exists, or non-performance of the AI in accordance with it. The second would be on the grounds of additional, relevant context which the AI was not developed to consider. A full 'rehearing' might only be appropriate where the potential damage to fundamental rights and economic interests is very serious – criminal justice AI decisions would be the obvious example.

8.6. Scope

To the extent that there is any global consensus about the scope of AI regulation, that consensus might be that regulation should be risk-based. Thus the UK's 2023 White Paper sets out a policy which aims 'to support innovation while providing a framework to ensure risks are identified and addressed … [and] to take a balanced approach to weighing up the benefits versus the potential risks.'[540] The EU's 2021 proposal for an AI Act describes its content as 'a proportionate regulatory system centred on a well-defined risk-based regulatory approach.'[541] The

[540] UK AI White Paper (n 435) foreword.
[541] EU proposal for an AI Act COM(2021) 206 final, Explanatory Memorandum, 3.

US National Institute of Standards and Technology has four mission objectives specific to AI, the second of which is to 'support the development of a risk-mitigation framework for deploying artificial intelligence systems.'[542]

Under a risk-based approach regulatory schemes are graduated so that high-risk activities receive the highest level of regulation, while at the low end of the risk scale there might be little or no regulation. This certainly seems to be the best approach for fairness regulation because the consequences of an unfair decision are not uniform. Society is likely to insist that an AI decision which might lead to imprisonment *must* be made fairly. We might contrast that with a decision about which passengers on an over-booked flight should be 'bumped' (if it were ever worth devising an AI to make that decision) – there is little or no social expectation that the airline should make that decision fairly, with some legal or regulatory sanction if they do not.[543]

Some AI decisions seem to raise no fairness considerations at all, and should thus be excluded from fairness regulation. For example, there seems no reason why the producers of a smart refrigerator, which monitors the food stored in it and tells the user what is no longer fit for consumption, and also re-orders items which are running low, should be required to consider any questions about the fairness of those decisions.[544] Similarly, fairness has nothing to do with the decisions an autonomous vehicle makes about its driving, which all focus on operational matters such as navigating the route and avoiding collisions.

A further reason why it would be inappropriate to try to regulate all AI for fairness is the difficulty in defining AI,[545] and the extent to which some kind of AI is soon likely to be built into a vast range of technologies. For example, much voice dictation technology is already AI-powered, together with content recommendation systems, online advertising, and so on. Imposing fairness requirements so widely would be a radical departure from the current social consensus about fairness regulation, which limits it to decisions which have a substantial impact on individual rights and interests.

I suggest, therefore, that AI fairness regulation needs a carefully defined scope. We have seen in the previous chapters that fairness questions arise when a decision affects an individual's fundamental rights and economic interests, and so it would seem sensible to limit regulation to technological systems which make such decisions autonomously.

[542] US National Institute of Standards and Technology Act (15 USC 271 et seq) section 22A.

[543] We might hope that the airline would, for example, be sympathetic to a passenger who was travelling to see a dying relative, but there is not currently any legal or regulatory obligation on the airline to do so.

[544] It might be thought that there is a fairness issue if the refrigerator's AI is biased towards the products of a particular manufacturer and thus will not order rival products. However, further consideration suggests that this is only problematic if the refrigerator producer built this bias into the AI, in which case it would better be dealt with as a competition/antitrust law issue. If the bias is learnt from the owner's purchasing preferences, that is hardly an unfair bias which regulation needs to address.

[545] See ch 1's note on terminology.

This then raises the further question – how far down the risk scale should regulation reach?

The starting point should be that an obligation to produce a risk mitigation explanation should be imposed only where the AI presents a serious enough risk. Deciding what risks are sufficiently serious to require regulation is not a question for theorists, but rather ought to reflect the social consensus which might be different in each country, reflecting its social norms and needs.

There are two ways in which lawmakers might approach this issue. The first is to set out in legislation or regulation a comprehensive list of the risks which fall within its scope. This is what the EU has done in its AI Act, under which high-risk AI systems are those that have a significant harmful impact on the health, safety and fundamental rights of natural persons.[546] That broad principle is turned into a long list in Annex III, which sets out uses of AI that range from the biometric identification of individuals, through assessment of student learning outcomes, employment selection and law enforcement, to explaining the law and its application to judicial authorities.

There is a clear problem with this approach, in that it requires a remarkably accurate level of ability to predict future uses of AI.[547] The danger of both under- and over-inclusion is obvious. As examples, our hypothetical AI which allocates scarce medical treatment seems to fall outside the list (unless the AI is considered to be evaluating the eligibility of individuals for public healthcare services,[548] which receives different national answers because whether medical treatment is a public service is defined by the structure of each EU Member State's health system and not by the nature of the AI decision). By contrast, applying the regulation to an in-class AI student assessment which does not affect students' final marks, as Annex III requires, seems excessive for what is a rather low-risk activity.

The second is to define risk in broad terms, such as 'serious' or 'substantial' risk to the fundamental rights and economic interests of those who are the subjects of an AI's decision. At first sight it seems as if this is unhelpful to AI producers – how are they to decide whether their AI system falls within the scope of the regulation? However, it must be remembered that the regulatory obligation we are discussing is to produce a risk mitigation explanation, which explains how the humans responsible for the AI's development, training, testing and implementation have evaluated foreseeable fairness risks and guarded against them. Undertaking this kind of risk assessment is a normal part of any technology development process, and because the range of uses to which AI will be put is huge and unpredictable it is only the AI producer who can identify in advance what the risks are, and thus how important each is. Concepts such as 'serious' or 'substantial' are familiar to

[546] EU AI Act (n 466), Art 6(2a).

[547] There is a mechanism for amending the Annex III list in Art 7, but the time required to consult on and agree changes means that amendments are likely to be infrequent.

[548] EU AI Act (n 466), Annex III, para 5(a).

humans in their daily lives, in the same way as the concept of reasonableness on which much law and regulation is currently based.

Of course, the ultimate decision on whether any particular risk is serious or substantial would be determined by the regulator, if regulatory sanctions are imposed, or by the court assessing a liability claim based on failure to take reasonable steps to produce a risk management explanation. This does leave some uncertainty as to the scope of the regulation, but in my view that uncertainty is acceptable. Once risks have been identified by an AI producer it will be clear that some fall within the regulation and some outside. At the border there will inevitably be an element of Hart's 'penumbra of doubt',[549] and AI producers will need to decide whether to include those risks in their risk mitigation explanation. In practice it is highly likely that where an AI producer is in doubt, they would decide for safety's sake to produce a risk mitigation explanation. Risk identification and analysis should already have been undertaken as part of the AI development and implementation process, and so the additional burden of explaining the results will be minimal.

8.7. Consequences of Non-compliance

Schemes of regulation tend to fail to achieve their aims if there are no adverse consequences for non-compliance. There are many ways to incentivise compliance, and so the suggestions here are not intended to be definitive – each jurisdiction has its own approach to regulatory compliance and its preferred schemes of sanctions and enforcement. What follows is based very much on the UK's approach to similar kinds of regulation and civil liability, but it might also suggest a starting point for other jurisdictions to consider.

The regulatory scheme proposed above has two elements:

- The obligation to use good faith and reasonable efforts to prepare a justified risk mitigation explanation and to publish it.
- The obligation to take reasonable steps to ensure that the AI decision system operates in accordance with that explanation.

There are two ways in which law and regulation might impose consequences for non-compliance:

- Regulatory action. This could be in the form of enforcement notices, requiring the non-compliance to be remedied, or the imposition of monetary penalties, or both. These are the two main powers of the ICO which it can use to enforce compliance with data protection law and the other privacy legislation which falls within its remit.[550] In the calendar year 2022 the ICO

[549] Hart (n 375) and accompanying quotation.
[550] For an explanation of when the ICO considers it appropriate to use these powers, see ICO (n 307).

issued 23 enforcement notices and imposed 33 monetary penalties. In the majority of cases a monetary penalty for past breaches was imposed simultaneously with an enforcement notice to prevent future breaches. In those cases where a monetary penalty was imposed without a simultaneous enforcement notice, either the regulatee had given an undertaking to comply for the future or was entering liquidation. It is worth noting that a monetary penalty can be imposed even if no prior enforcement notice was issued, and also that enforcement notices are usually complied with because failure is likely to lead to a monetary penalty.

- Civil liability. The obligation to produce a risk mitigation statement is, at heart, an obligation to take reasonable care to avoid damaging the fundamental rights and economic interests of decision subjects. This is closely analogous to the duty under the tort of negligence to take reasonable care to avoid foreseeable risks of loss or damage to others. Imposition of a duty of care in negligence requires foreseeability that the person in question might suffer loss or damage, and here the AI user (who has the primary duty) knows directly that the decision subject is at risk because the user has chosen to use the AI to make a decision about that subject. Negligence leads to an obligation to compensate the victim, and I suggest that where the breach of my proposed regulatory fairness obligation causes loss to an individual that individual should have a similar compensation claim against the AI producer.

The first element of the regulatory obligation might be contravened either by: lack of good faith, such as making no serious effort to identify risks and devise ways to mitigate them; or failure to use reasonable efforts to produce the explanation, in other words that the AI producer genuinely tried but failed because, for example, inexperienced staff were allocated the task and were unable to carry it out properly. Here the regulator would have the power to issue enforcement notices and/or monetary penalties if it was satisfied, after investigation, that the AI producer was in breach.

For a civil liability claim, the decision subject would need to prove lack of good faith or failure to take reasonable care. In the most egregious cases the compliance failure would be obvious on the face of the explanation itself, so that little or no additional evidence would be needed to persuade a court of the breach. In less obvious cases the claimant decision subject would need to prove lack of good faith or failure to take reasonable steps in the normal way, primarily through discovery of the internal documents and communications of the AI producer which relate to the production of the explanation. It is important to note that the issue to be proved here is a failure of required human behaviour, and that no investigation is likely to be needed into the technical workings of the AI. Imposing the burden of proof on the claimant thus seems a reasonable solution.

The second element, that the AI producer has not taken reasonable care to ensure that the AI is operating in accordance with the risk mitigation explanation, is more complicated. If the burden of proof were placed on the regulator and the claimant decision subject, it might be impossible to discharge without lengthy and expensive expert investigation of the AI's workings. I suggest, therefore, that in both cases the regulator or the claimant should merely have to raise a reasonable suspicion that there has been a failure in this respect. That suspicion is likely to arise from the fairness mitigation explanation itself and perhaps from relatively limited discovery. If the explanation for a medical treatment allocation AI claims, for example, that appropriate steps have been taken to eliminate bias against female applicants, one route to raising such a suspicion to found a claim by a woman who was rejected would be via a statement from her doctors that they believed she met all the criteria set out in the explanation. Alternatively she might seek to discover the proportion of men and women selected by the AI, which might also raise such a suspicion. The sectoral regulator in this instance might well impose a requirement to capture and publish statistics about the AI's performance, and these would clearly be useful to both the regulator and to claimants.

Suspicion itself is not, of course, enough to determine a regulatory breach or a right to compensation. However, all the information about how the AI performs in relation to the risk mitigation explanation is solely in the possession of the AI producer. This justifies a reversed burden of proof – once reasonable suspicion is raised, it will be for the AI producer to demonstrate that reasonable care was in fact taken.

Alternatively, both regulatory liability and civil liability might be strict for the second element. Failure to operate in accordance with the risk mitigation explanation is arguably careless by definition, because the risk mitigation explanation defines the promised standard of care. If strict liability were thought more appropriate, it might still be useful to provide a defence that the AI producer had taken all reasonable efforts to avoid the failure. Given the inevitable uncertainties about how an AI will decide an individual case, the AI community would no doubt welcome such a defence. However, it can also be argued that the cost of compensating those to whom the AI causes loss ought to be a cost of implementing and using the AI in the first place. If so, civil liability might be strict, but regulatory liability be limited to an enforcement notice requiring the failure to remedied for the future.

8.8. A General AI Regulator?

Section 8.4.2 explains how sectoral regulators would be likely to expand on the basic obligation to produce a risk mitigation explanation and ensure that the AI operates in accordance with it. This leaves us with the question whether it is sufficient to establish the basic obligation in the general law, or whether it needs a general AI regulator to oversee and manage its future development.

Existing sectoral regulators are well-placed to deal with AI uses which clearly fall within their sectors. However, AI will be implemented in ways which will not clearly fit in those sectors, while still generating risks. Even though the basic fairness obligation would apply to new AI applications which create serious risks, AI producers might find it hard to meet that obligation without authoritative guidance.

Whether a general regulator is needed depends on the decision which is made about the consequences for breach of the general obligation. If the only consequences are civil liability, then the role of a new, general AI fairness regulator would be limited to issuing guidance. Establishing a regulator to perform this function alone seems excessive, and the regulation could simply mandate the courts to consider any sources of external guidance which are relevant. Alternatively, some coordinating function could be established to bring together recommendations from external bodies and experts, as proposed by the AI White Paper.[551]

If, though, it is thought that enforcement notices and monetary penalties are required to secure compliance in appropriate cases, then a general AI fairness regulator would need to be established. Its role would encompass both guidance and enforcement, and the legislation establishing the regulator would need to define the extent of its powers to issue enforcement notices and impose penalties. Each state has its own view about how regulatory compliance is best achieved – as an example, it is generally recognised within the data protection community that the UK regulator focuses more on securing future compliance and less on punishing breach, while the French regulator has a much stronger focus on breach and punishment. An important consideration for all states, though, must be to avoid imposing such a high regulatory burden on the developing AI sector that it discourages potentially useful AI implementations.

A definition of powers which might fit well with the UK approach to regulation would be that the regulator may:

- Issue guidance on compliance with the general fairness obligation.
- Issue enforcement notices where the regulator considers that:
  - a fairness mitigation explanation is defective and creates a serious risk of damage to the fundamental rights and economic interests of decision subjects; or
  - an AI decision system has failed to operate in accordance with its fairness mitigation explanation and has produced unfair decisions which resulted in serious damage to the fundamental rights and economic interests of decision subjects.
- Impose monetary penalties where the regulator considers that AI user has failed to act in good faith and use reasonable efforts when producing its risk mitigation explanation, with the level of penalty bearing some relation to the adverse consequences which have resulted or might result.

[551] UK White Paper (n 435), section 4.3.1.

For a regulatory system comprised of both general and sectoral regulators to work optimally there would need to be a high level of communication between the general regulator and the sectoral regulators, and between sectoral regulators. This coordinating function could be performed by the general regulator. As an example, fairness risks and mitigation strategies identified in the medical sector would potentially be of interest to the financial sector (and vice versa).[552] Sectoral regulators would develop deep understandings of AI which might suggest improvements to and extensions of the general guidance and regulation. Because AI technology is often not sector-specific, a joined-up system is important.

It also seems likely that, as AI is implemented widely across all areas of human life, we will identify new sectors of activity which need regulating. The Online Safety Act 2023 has already, in effect, decided that social media needs to become a new regulated sector, although this is achieved by extending the remit of an existing regulator rather than setting up a new one. Society looks for fairness everywhere, and the adoption of AI decision-making might identify that some uses create new fairness risks which are serious enough to require regulatory control. Identifying these new sectors is an obvious task for my proposed general regulator, through making proposals for consideration by the legislature rather than taking on that regulatory role itself.

8.9. Application in Practice

To illustrate how the regulatory system proposed above might work in practice, this section of the chapter sets out three examples, all assumed to be AI systems used in the UK. For each, I suggest ways in which regulatory failures might occur and the consequences which might follow that failure. For ease of comparison between them, each of the AI systems will be used for recruitment selection, choosing applicants to be shortlisted for interview and rejecting all other applicants. No applicant will complain about being selected unfairly, and so the analysis will focus on those applicants who might have been unfairly rejected.

8.9.1. Bespoke AI

In this example, a large employer (the AI user) commissions the AI developer to produce a recruitment selection AI. The employer provides a set of merit criteria for each job for which the system will make a selection, and makes its records of past job advertisements and applicant CVs available to the developer. From those records the developer will produce a set of training data for the AI.

[552] As a non-AI example, the aviation safety practice of working to checklists is (after some opposition) being introduced in medicine – see Robyn Clay-Williams and Lacey Colligan, 'Back to basics: checklists in aviation and healthcare' (2015) 24 *BMJ Quality & Safety* 428.

Because the regulatory system is in force, the employer and the developer discuss how fairness issues should be addressed in the development of the AI. They review the guidance issued by the general regulator, which sets out two specific recommendations for recruitment selection systems:

- the merit criteria for selection should not include criteria which do not relate to performance of the job duties;
- unlawful discrimination should be avoided.

They agree that the developer will use its expertise and understanding of unlawful discrimination to address the second, while the employer will take responsibility for the first.

The system is developed, trained and tested and appears satisfactory, a fairness mitigation explanation is written and posted on the employer's company website, and the system is put into operation. However, after a few months the employer receives notice of a complaint made to the general regulator by a dissatisfied applicant for a management position. The applicant claims that they were rejected by the system, but that another candidate with lower qualifications was accepted for interview, and that the applicant suspects that the reason was that the accepted candidate was educated at a fee-paying school while the rejected candidate went to a state school. After some investigation, the employer realises that 90 per cent of its current and past managerial employees were educated at fee-paying schools, and that the place of education was not considered as a possible fairness issue during the AI development. It is likely that the machine learning process identified this criterion as an important characteristic for applicants. Selection on an unstated basis which is not relevant to performing the job, in this case place of education, is clearly likely to be seen as unfair.

The first question to be decided is whether there has been a breach of the general fairness obligation, which requires good faith and reasonable efforts to identify serious fairness risks and to mitigate them. The good faith of the employer, who owes this duty, is not in question. Similarly, the developer has acted in good faith. The question is thus whether the risk that selection is made partly on the basis of place of education presents a serious risk to the fundamental rights and economic interests of applicants, and if so whether the employer made reasonable efforts to identify and mitigate it.

Place of education is not a protected characteristic under section 4 of the Equality Act 2010, so no unlawful discrimination has taken place. In the UK there is no fundamental human right to employment, so the question is whether the applicant has suffered damage to their economic interests. Clearly there is some damage, which is the loss of an opportunity to be considered at interview for the management position. However, this is the loss of a chance only, not of a right to the position, and the applicant might well not have been successful at interview for a range of reasons, including that the employer did not like them as much as other well-qualified candidates. It is hard to see this risk as a serious one, and so the employer will not be in breach of the general fairness obligation by not identifying and guarding against it. Thus there should be no right to compensation for the applicant, and no regulatory consequences.

If, though, we add in that the employer is in the UK top 10 list by numbers employed, the general regulator might take the view that allowing the employer to continue recruitment using a system which gives excessive weight to the place of education would contravene the broader aims of the regulation, because of the number of UK residents potentially affected, and thus constitute a serious fairness risk. If so, it might seek an undertaking from the employer that the system would be modified to deal with this bias, and issue an enforcement notice if the employer refused. An appeal against such a notice would lie to the courts, as is the case for enforcement notices under data protection law.[553] Ultimately, it will be for that appeal body to interpret the regulation and decide whether it encompasses this type of unfairness as a serious risk for which there is a regulatory power to engage in enforcement action.

8.9.2. Customisable AIaaS System

Here we have a medium-sized UK employer, the AI user, who cannot justify the cost of a fully customised recruitment selection system. The AI developer, a US corporation, offers use of its recruitment selection system on an AIaaS basis, with users paying an annual subscription and a per-applicant fee. The system allows users to input their chosen merit criteria, in terms of qualifications, skills and experience, and also has a free-form input for 'Other Employer Preferences' which is used to retrain the AI to meet the employer's requirements. The AI developer's website has a page headed 'Legal Compliance' which explains that the system has been designed to avoid unlawful discrimination and bias in accordance with the guidance given by the US Equal Employment Opportunity Commission (EEOC).[554] The employer signs up to use the recruitment system, and its compliance manager produces a web page which sets out a fairness risk mitigation explanation which makes two statements:

- the employer sets the merit criteria for selection in order to select the candidates who are the best fit for each position; and
- the selection system's developer has ensured legal compliance, and provides a link to the relevant AI developer's web page.

The employer advertises for an entry-level position, and because it is concerned that its workforce is older than normal for that sector writes 'Age under 25 preferred' in the 'Other Employer Preferences' input box. This produces an

[553] Data Protection Act 2018, s 162.
[554] www.eeoc.gov/prohibited-employment-policiespractices.

unusual consequence – because US anti-discrimination law prohibits discrimination against individuals over 40,[555] the recruitment selection system automatically rejects every candidate between the ages of 25 and 39.

This amounts to unlawful age discrimination under the Equality Act 2010 because that Act makes all discrimination on age grounds unlawful, with no cut-off limit specified. Is there also an infringement of the UK's fairness regulation?

I suggest there clearly is. In its fairness mitigation explanation the employer has made no effort at all to avoid unfair bias, simply relying on the developer's 'Legal Compliance' web page. It has not identified the risk that US law might differ from UK law, and in explaining the merit criteria it has not given any thought to whether those criteria are fair in relation to the needs of the job – age seems to be irrelevant to performance, and the employer is seeking a younger work force for unrelated reasons. The employer acted in good faith, but has not made reasonable efforts to identify fairness risks and to mitigate them. If the risk of age discrimination had been considered and investigated, it should have been clear that the AI developer's system was not designed to deal with that risk in accordance with UK law.

Here, the regulatory consequences might include a monetary penalty, if the regulator considers the breach sufficiently serious, together with an enforcement notice. A compensatory claim by rejected applicants aged 25–39 would also be available, but because each candidate's loss is speculative a court would likely only award nominal damages.

8.9.3. Generative AI

In this example we have a small employer who is time-poor and has no real understanding of AI. Open AI's ChatGPT is free to use, so the employer decides to save effort by using it to select interviewees from the applicants for a job. After some cursory reading to understand how to write prompts for ChatGPT, the employer sets out the qualifications, skills and experience which are looked for and asks ChatGPT to rank each applicant CV in numerical order from best to worst. The employer is unaware of the UK's AI fairness regulation, and so does not produce a risk mitigation explanation.

ChatGPT, with its origins in large language models, is designed to emulate a human conversationalist by providing the most likely response to its input, based on its training on a large proportion of the freely available online content. It was not designed to make decisions about the subject of conversation, in this case the best potential recruit, and its responses are simply those which it assesses as most likely from the data on which it was trained.

[555] ibid.

Consequently, although the rankings which ChatGPT produces have some relationship to applicant qualifications, skills and experience, its responses are also influenced by the biases which humans have exhibited in their online content.[556] Additionally, as OpenAI notes, ChatGPT can produce 'plausible-sounding but incorrect or nonsensical answers',[557] also known as 'hallucinations'. The ranking list therefore successfully rejects most clearly unqualified candidates, but its ranking of those who are qualified seems random and without any rational basis. On any human understanding of the concept, the selection has undoubtedly been unfair.

A breach of the UK fairness regulation has clearly occurred. The employer has not considered producing a risk mitigation explanation, and has made no attempt to even consider the potential risks of using ChatGPT in this way. If the risks had been considered, it should have been clear that there was no way of mitigating them – ChatGPT is simply unsuitable for the task.

Given the small size of the employer, the regulator might decide simply to issue an enforcement notice preventing future recruitment selection in this way. Compensation claims by individuals would theoretically be possible, as in the previous example, but probably not worth pursuing.

More importantly, this example will have told the general regulator that small businesses can easily be tempted to use AI tools, especially those which are free, for purposes for which they were not designed and for which they are wholly unsuitable. It would therefore likely issue warnings and guidance, publicising them as widely as possible, and also liaising with bodies like the Equal Opportunities Commission to ensure that their own guidance was updated to cover this issue. This guidance, warning and liaison function of the regulator will probably do more to achieve AI fairness in the long term than any individual enforcement action might achieve.

8.10. Fairness and Wider AI Regulation

The idea of establishing a general AI fairness regulator is perhaps a little idealistic, at least at the time of writing. Fairness is important to society as an aspiration for decision-making, but in practice we regularly experience unfair decisions without expecting law and regulation to remedy them.

I suspect, though I cannot prove, that this everyday level of unfairness is acceptable to society because human decision-makers usually try to act fairly, even when they are under no obligation to do so. Society's fairness norms are

[556] Partha Pratim Ray, 'ChatGPT: A comprehensive review on background, applications, key challenges, bias, ethics, limitations and future scope' (2023) 3 *Internet of Things and Cyber-Physical Systems* 121.
[557] openai.com/blog/chatgpt.

very strong, and social norms influence our behaviour at least as strongly as legal norms do.

However, AI systems are immune to human social norms unless compliance with them is built in. As AI takes an increasing number of decisions away from humans, there is an obvious risk that the ground level of unfairness might increase to a socially unacceptable level.

For this reason, I suggest that states should seriously consider introducing a general obligation of fairness for those AI producers whose AI systems pose serious risks to the fundamental rights and economic interests of decision subjects. The obligation would be to produce and publish a risk mitigation explanation, and the sanction for breach would be civil liability to the decision subject for failure to use good faith and reasonable care to produce the explanation, or where reasonable efforts were not made to ensure that the AI performed in accordance with the explanation. Producing such an explanation requires little more than good practice already suggests, and the risk of liability is large enough to incentivise compliance but not so large as to disincentivise developing and using the AI.

Sectoral regulators will, where they are not already doing so, concern themselves with fair AI decisions. The regulatory model proposed above is one which they could easily adopt, at least as an interim measure until the full extent of fairness risks in their sector becomes apparent and demands additional obligations on AI producers.

In the longer term, though, I expect that the other risks which AI generates will, cumulatively, be seen as sufficiently great that some general AI regulator will be needed. If this happens, regulating for fairness will be an obvious part of such a regulator's remit.

There are already signs that a need for general AI regulation may soon be recognised. At the May 2024 AI Seoul Summit sixteen AI technology companies, including Amazon, Google, Meta and Microsoft, signed up to the *Frontier AI Safety Commitments*.[558] They voluntarily commit to undertake risk assessments of their frontier AI models and systems, both during development and post-deployment, to attempt to mitigate identified risks, and not to deploy AIs whose risks are too serious. They also agree to explain to both the public and to governments how they have done this, which implicitly requires them to justify their actions.

These commitments match very closely the concept of justified risk mitigation explanations set out in this chapter. To develop them into a regulatory system the obligation would need to be imposed by law, perhaps setting out in regulation specific risks which should always be considered rather than leaving the AI producer with a free choice. An obligation for the AI operate in accordance with its explanations would also be desirable. Consequences for breach, including failure

[558] www.gov.uk/government/publications/frontier-ai-safety-commitments-ai-seoul-summit-2024/frontier-ai-safety-commitments-ai-seoul-summit-2024.

to do these things adequately, would be necessary. A state which took this route would necessarily need some regulatory body to operate the regulatory system, and as the regulation covers a wide range of risks, not merely that of unfair decisions, a general AI regulator would come into existence.

Many of the risks which a general regulator might be expected to deal with share important characteristics with fairness. In the following chapter I will briefly examine three of these – consistent functionality, explainability and trustworthiness. All these exhibit similar problems to those we have found with regulating fairness:

- They are not absolute obligations. Just as we want AI to be fair enough for social acceptance, we also want AI to be ethical, consistent, explainable and trustworthy enough. Our social understanding of 'enough' changes over time, and we want AI to keep up with it.
- They are equally as hard to define as is fairness. We can easily recognise that an AI is untrustworthy, for example, but find it very hard to say what would make it trustworthy or even that it is actually trustworthy.
- Ultimately, all three are emotional responses to the AI. The regulatory aim is that we *feel* that the AI is adequately ethical, consistent, understood and trustworthy. Regulating to achieve the emotional response of fairness is what this book has been about, and so it seems likely that the regulatory model proposed in this chapter might be adaptable to regulate these other risks.

There may also be other risks generated by AI which, on analysis, can only be mitigated by engendering an emotional response in society. I conclude the book by examining the issues of generative AI which is trained on intellectual property belonging to others. Scholars have already identified that the concept of intellectual property infringement does not deal with these issues adequately, if at all, and it seems worthwhile to ask if my regulatory model might be helpful here.

9

Beyond Fairness

In this final chapter I will investigate how far the regulatory tool of justified risk mitigation explanations might be useful for achieving aspirational values for AI which we cannot define and specify well enough for traditional modes of regulation to be useful. As we will see, the question of whether there has been compliance with these aspirational values seem to have a strong emotional element, just as fairness does. There are many such values and insufficient space in this chapter to investigate all of them. I therefore propose to concentrate on three of them:

- consistent functionality;
- trustworthiness; and
- explainability.

My hope is that these are sufficiently representative that we can draw useful conclusions about the utility of justified risk mitigation explanations as at least part of the regulatory toolbox.

This does not mean that I am proposing that all, or even some of, these values, should be imposed on every AI producer through regulation. The decision to regulate is a complex one, which needs to weigh the potential benefits to society against the costs. For many uses of AI, minimal or even no regulation will turn out to be the best answer once that cost–benefit assessment has been made.

In general, the answer to whether regulation is desirable will depend, just as it does for fairness, on how seriously a failure on the part of the AI to comply with the value in question is likely to affect the fundamental rights and economic interests of individuals who are impacted by the AI's activities. If we think back to our example of an AI which allocates scarce medical treatment, we will certainly want it to make consistent recommendations which are trustworthy. However, that AI is operating as part of the system of wider medical diagnosis and treatment, and the system as a whole contains safeguards which mean we might, initially at least, trust the medical staff to ensure that these values are complied with. It would only be if the AI effectively displaced the judgement of those medical professionals that we would need to regulate the AI producer to ensure that consistency and trustworthiness were embedded in the AI. At the other end of the spectrum, an AI which makes recommendations about the music I might like to listen to needs no regulation in those respects. If its recommendations are inconsistent, untrustworthy and inexplicable I will simply stop using it.

Finally, I conclude the chapter by asking whether justified risk mitigation explanations might have a use beyond aspirational values, as a first step towards identifying the solution to social conflicts caused by AI and whose resolution is not obvious. Generative AI has generated such a conflict, between producers of intellectual creations and the AI producers who use those creations to train their generative AIs. This activity is so different from previous uses of intellectual creations that the area of law designed to deal with such conflicts, intellectual property, seems likely to offer few meaningful answers. Requiring generative AI producers to identify the new risks to the fundamental rights and economic interests of creators, and explain how they have mitigated those risks, has the potential to identify the real issues at stake and help to outline an appropriate social settlement which could be embodied in future reform of intellectual property law.

9.1. Ethical AI

Before we start, it is worth asking whether we might wish to use the regulatory tool of justified risk mitigation explanations to ensure ethical AI in its broad sense. We saw in section 4.1.1 that the work of ethicists has identified a number of principles which AI ought to comply with. Fairness is just one of these. All those ethical principles specify in very broad terms the outcome which AI is desired to achieve, but as explained earlier, offer little insight into how law and regulation might mandate compliance with them, or at least increase the likelihood that they are complied with.

A useful summary of the ethical values most commonly proposed for AI can be found in the list of 11 principles which Singapore has identified from regulatory and ethical initiatives around the world:[559]

- transparency;
- explainability;
- repeatability/reproducibility;
- safety;
- security;

[559] These are checked as part of its AI Verify initiative – Singapore Infocomm Media Development Authority (IMDA), *Fact Sheet – Open-Sourcing of AI Verify and Set Up of AI Verify Foundation* (2023), www.imda.gov.sg/-/media/imda/files/news-and-events/media-room/media-releases/2023/06/7-jun--ai-annoucements---annex-a.pdf. The initiative aims to develop 'AI testing frameworks, code base, standards, and best practices', some of which will be in the form of open source code software tools, while others will consist of process checks.

There is, of course, a risk that AI producers whose AIs pass these checks will think that their work is done, and that their AI is ethically compliant. The IMDA recognises this risk, making it clear that 'AI Verify does not set ethical standards, neither does it guarantee AI systems tested will be completely safe or be free from risks or biases.' It remains to be seen how AI Verify will be integrated into Singapore's wider regulatory system for AI, and how this risk will be managed.

- robustness;
- fairness;
- data governance;
- accountability;
- human agency and oversight; and
- inclusive growth, social and environmental well-being.

Outcomes for most of these ethical principles are not definable in any coherent way, because the desired outcome shares with fairness the characteristic that it is primarily an emotion generated in society. Security is not an absolute value, for example, because there is a continuing arms race between those who want to overcome security measures and those who wish to maintain them. Security is a moving target, and the real question is whether a technology is secure enough to achieve acceptance from those who are its stakeholders: users of the technology, regulators and ultimately society as a whole. Transparency requires sufficiently full and clear communication to its intended audience so that they feel satisfied that they understand the workings of the technology well enough to accept its use. The ultimate compliance test for all these principles is whether stakeholders are happy or content (both emotions) that the technology accords with the principle.

This means that rule-based law and regulation can only go part of the way to securing compliance with ethical principles for AI. The final hurdle is to secure social acceptance of the AI, based on emotional response. This is exactly the issue we have been examining with respect to fairness, which suggests that regulatory requirements for justified explanations might be a useful tool to help overcome that final hurdle. We have seen when examining fairness that these explanations can be useful where there is no clear or right answer about how AI producers should behave to achieve the desired outcome, and where choices need to be made between different techniques to do so.

Regulation which demanded such a broad spectrum of ethical compliance and imposed sanctions for failure would clearly be excessive. Each AI implementation creates different ethical risks, some of which are more serious than others. Thus although there might be social value in requiring AI producers to consider those ethical risks and explain how they have dealt with them, the decision whether to impose some legal or regulatory consequences for particular ethical failures is best left to regulation at the sectoral level.

9.2. Consistent Functionality

It seems uncontroversial that an AI which is implemented to carry out a task should be expected actually to do what it is supposed to do, and should do so consistently. Consistency in decision-making is, as we saw earlier, an important element of the human conception of fairness. The core of fairness is that like cases should be treated alike. Even where fairness is not at issue, society is likely to look for consistency in other activities which AIs control.

As just one example, generative large language model AIs are being linked to search engines, so that responses to searches can be given as textual answers rather than just by links to sources which might offer answers. Such a search engine would be less useful, perhaps even useless, if it gave a materially different answer every time the same question was asked.

Although it might not, yet,[560] be socially necessary to regulate search engines for consistency, we can easily envisage AI applications where inconsistent outputs might create serious risks. An example might be an AI which advises engineers working on power lines about the safety precautions which they should take, customised for the particular incident they are working on. Inconsistent advice, some of which would therefore be wrong, is clearly confusing at best and dangerous at worst.

How to implement the ethical principles of repeatability/reproducibility and robustness are active topics in the AI research community, as we will see below. Society needs to be assured that AI which affects the lives of its members is working consistently, rather than making capricious decisions. Predictability in machine decision-making is just as socially desirable as predictability in human decision-making. Consistency captures both of those ethical principles, and so here I attempt to explain how justified explanations might be used to satisfy stakeholders in that respect without offering them metrics. Current research into these matters aims to show consistency by testing against objective metrics, but the mere fact those metrics have been met will not be enough to convince wider society because most individuals are not emotionally satisfied by metrics alone.[561]

Raji et al argue that inconsistent functionality is an inevitable result of the false belief that AI systems actually do what they are claimed to do:[562]

> Despite the current public fervor over the great potential of AI, many deployed algorithmic products do not work. AI-enabled moderation tools regularly flag safe content, teacher assessment tools mark star instructors to be fired, hospital bed assignment

[560] At least at present. Recent research finds that once the output of generative AI enters the training dataset then the process of 'model collapse begins', in which the output of a generative AI trained on that dataset becomes increasingly devoid of accuracy and tends towards nonsensical statements – Ilia Shumailov et al, 'The Curse of Recursion: Training on Generated Data Makes Models Forget' (2023) *arxiv preprint, arXiv:2305.17493v2*. That would potentially create a serious risk for those relying on search results if inaccurate answers might be relied on and cause loss or damage.

[561] Only scientific training appears to change this. A few years ago I asked an audience of lawyers, regulators and policymakers whether they would be happy as pedestrians to run the risk of being hit by a self-driving car, as opposed to a human driver, assuming the self-driving car had been demonstrated to have substantially fewer accidents. There was near unanimity that human drivers were preferable, even though some audience members recognised this was irrational but without knowing why. A few months later I posed the same question at the Royal Society in London, where there was unanimity that the self-driving car was a far better risk.

[562] Inioluwa Deborah Raji, I Elizabeth Kumar, Aaron Horowitz and Andrew Selbst, 'The fallacy of AI functionality' (2022) *Proceedings of the 2022 ACM Conference on Fairness, Accountability, and Transparency* 959 (references omitted). See also at 960:

algorithms prioritize healthy over sick patients, and medical insurance service distribution and pricing systems gatekeep necessary care-taking resources. Deployed AI enabled clinical support tools misallocate prescriptions, misread medical images, and misdiagnose. The New York MTA's pilot of facial recognition had a reported 100% error rate, yet the program moved forward anyway. Some of these failures have already proven to disproportionately impact some more than others: moderation tool glitches target minoritized groups; facial recognition tools fail on darker skinned female faces; a hospital resource allocation algorithm's misjudgements will mostly impact Black and lower income patients. However, all failures in sum reveal a broader pattern of a market saturated with dysfunctional, deployed AI products.

Their in-depth analysis of the reasons for such functionality failure identifies matters such as:

- where the AI relies on data or proxies which have no causal link to outcomes;[563]
- where training data is inadequate because it contains errors or is incomplete;
- where testing was not undertaken properly or thoroughly; and
- where the AI is used outside the domain in which it has actually been demonstrated to work without testing whether it works in the new domain.

The question is therefore whether a requirement to provide risk mitigation explanations might be a useful way to deal with consistency issues. Is there a 'right' answer about how to achieve consistency? If not, would explaining and justifying the AI producer's choices about risk mitigation be likely to reduce the risk of inconsistency?

The AI research community has considered the consistent functionality question, though not under that name. Discussions focus around a range of terms, the most common of which are repeatability, replicability and reproducibility.[564] Gundersen and Kjensmo explain the differences helpfully:

> R1: Experiment Reproducible. The results of an experiment are experiment reproducible when the execution of the same implementation of an AI method produces the same results when executed on the same data. This is often called repeatability.

The functionality of AI systems is rarely explicitly mentioned in AI principle statements, policy proposals and AI ethics guidelines. In a recent review of the landscape of AI ethics guidelines, Jobin et al found that few acknowledge the possibility of AI not working as advertised.

[563] This also potentially creates legal problems – see section 5.2.

[564] The terminology, and suggested meaning of, each term is complex – see Odd Erik Gundersen, Saeid Shamsaliei and Richard Juul Isdahl, 'Do machine learning platforms provide out-of-the-box reproducibility?' (2022) 126 *Future Generation Computer Systems* 34, 35 (references omitted):

> Drummond argues that replication means to exactly replicate the original experiment and that reproducibility is obtaining the same results from quite a different experiment. Stodden states that replication is re-running the experiment with code and data provided by the author, while reproduction is a broader term that implies both replication and the regeneration of findings with at least some independence from the original code and/or data. Peng suggests that reproducibility is a continuous variable ranging from only a paper describing an experiment being shared to the linked executable code and data being shared along with the paper. Goodman et al. present three different terms describing reproducibility: (1) Methods reproducibility means that the exact same procedures could be exactly followed, (2) Results reproducibility refers to obtaining the same

R2: Data Reproducible. The results of an experiment are data reproducible when an experiment is conducted that executes an alternative implementation of the AI method that produces the same results when executed on the same data. This is often called replicability.

R3: Method Reproducible. The results of an experiment are method reproducible when the execution of an alternative implementation of the AI method produces consistent results when executed on different data. This is often called reproducibility.[565]

Repeatablity is easily tested for by undertaking more than one pass of the same dataset. For example, when an AI is being tested the results of multiple applications of it to the same dataset can easily be compared, and ought to be identical in outcome for each case.

Replicability is more of a challenge. To test for it fully, AI developers would need to construct a new version of the AI using a different platform[566] or technology. This is expensive and time-consuming, and so it seems likely that developers will undertake more limited testing to assure themselves that their AI provides adequate replicability.

Reproducibility is what society is most likely to demand, and is the most difficult to demonstrate. Consistency, like fairness, demands that like cases be treated alike. By this, society does not mean identical cases, but rather that similar cases should receive similar treatment. In other words, what society needs to be assured of is that the socially irrelevant data relating to two cases which would be seen as similar by humans do not lead to completely dissimilar outcomes for those

results from conducting an independent study whose procedures closely match the original study and (3) Inferential reproducibility in which qualitatively the same conclusions can be drawn from an independent study or reanalysis of the original study. Gundersen and Kjensmo propose that for AI research three reproducibility degrees can be defined based on which documentation the original researchers share with independent researchers. The documentation could be divided into (i) the scientific report, (ii) the data and (iii) the code from the original experiment. Tatman et al. suggest three levels also based on what is shared: (1) Low reproducibility: paper is shared, (2) Medium reproducibility: paper, code and data are shared, (3) High reproducibility: paper, code, data and environment is shared. In the report by the National Academies of Sciences, Engineering and Medicine, reproducibility is defined to mean computational reproducibility and replicability to mean obtaining consistent results across studies answering the same scientific questions. While reproducibility does not necessarily mean discovery of truth, as Devezer et al suggest, enabling reproducibility makes the analyses and evaluations transparent. Plesser provides a historical overview while Gundersen provides a systematic survey of reproducibility definitions.

[565] Odd Erik Gundersen and Sigbjørn Kjensmo, 'State of the Art: Reproducibility in artificial intelligence' (2018) 32 *Proceedings of the AAAI Conference on Artificial Intelligence* 1644, 1647.

The terminological confusion is apparent from the fact that other authors swap the meaning of the two latter terms, eg: 'Reproducible work consists of re-doing an experiment using the same data and same analytical tools, whereas Replicable work considers different data (presumably sampled from similar distribution or method).' Joelle Pineau, Philippe Vincent-Lamarre, Koustuv Sinha, Vincent Larivière, Alina Beygelzimer, Florence d'Alché-Buc, Emily Fox and Hugo Larochelle, 'Improving reproducibility in machine learning research (a report from the NeurIPS 2019 reproducibility program)' (2021) 22 *Journal of Machine Learning Research* 7459, 7462.

[566] Much AI development is undertaken on third party platforms. For a description of the most common see Gundersen et al (n 564) 38.

cases. What follows will therefore concentrate on reproducibility as defined above, though the same issues also arise for replicability.

Methods for assessing reproducibility seem to be reasonably well understood, and the various explanations of these methods are in broad agreement.[567] However, this does not mean that regulation could prescribe when, and to what extent, those methods should be used, because in practice only some methods will be available, and not all will be affordable. Pineau et al summarise the problems as follows:[568]

> Compared to other fields like biology, physics or sociology where experiments are made in the natural or social world, the reliability and reproducibility of experiments in computer science, where the experimental apparatus for the most part consists of a computer designed and built by humans, should be much easier to achieve. Yet in a surprisingly large number of instances, researchers have had difficulty reproducing the work of others. Focusing more narrowly on machine learning research, where most often the experiment consists of training a model to learn to make predictions from observed data, the reasons for this gap are numerous and include:
>
> - Lack of access to the same training data/differences in data distribution;
> - Misspecification or under-specification of the model or training procedure;
> - Lack of availability of the code necessary to run the experiments, or errors in the code;
> - Under-specification of the metrics used to report results;
> - Improper use of statistics to analyze results, such as claiming significance without proper statistical testing or using the wrong statistic test;
> - Selective reporting of results and ignoring the danger of adaptive overfitting;
> - Over-claiming of the results, by drawing conclusions that go beyond the evidence presented (e.g. insufficient number of experiments, mismatch between hypothesis & claim).

For machine learning, a particular obstacle to assessing reproducibility is availability of the technical facilities needed. Gundersen at al[569] have reviewed the most common machine learning platforms which are used to develop AI, and conclude that each provides a different mix from the set of functionalities which are needed to assess reproducibility. No platform provides a full set of functionalities, and thus:

> Even though there exist good solutions for covering each of the variables described by the framework for assessing the reproducibility of machine learning platforms … these solutions are not provided to users of the assessed machine learning platforms. Hence, the full reproducibility potential is not enabled for experiments conducted on

[567] See eg the NeurIPS Machine Learning Reproducibility Checklist set out in Pineau et al (n 564) 7478; Odd Erik Gundersen, Yolanda Gil and David Aha, 'Towards reproducible research, open science, and digital scholarship in AI publications' (2018) 39 *AI Magazine* 56; Benjamin J Heil, Michael M Hoffman, Florian Markowetz, Su-In Lee, Casey S Greene and Stephanie C Hicks, 'Reproducibility standards for machine learning in the life sciences' (2021) 18 *Nature Methods* 1132.

[568] Pineau et al (n 565) 7460.

[569] Gundersen at al (n 567).

the platforms we surveyed. Researchers have to take extra steps and put extra effort into making their research reproducible.

The 'gold standard' for assessing reproducibility is if independent researchers can replicate the original findings. For any particular AI system, this would mean making the methods, code and data available to those independent researchers. This is not as simple as it sounds, for three reasons:[570]

- First and most obvious, if the AI has been produced for a commercial purpose, disclosing the information, code and data increases the risk that legal protections of confidential information will be lost. Confidentiality agreements with the independent assessors of reproducibility can reduce this risk, but the wider the confidential information is known, the higher the chance that it is disclosed. Additionally, academic researchers are the most obvious source of such assessment, but confidentiality obligations do not sit comfortably with the academic imperative to share and publish information.
- Second, the data on which an AI is trained and tested may itself contain confidential information or personal data. The terms under which that data was obtained are likely to prevent its disclosure to third parties, and data protection law may restrict the disclosure of elements which include personal data.
- Third, parts of the code may be licensed from third parties, and independent assessors may require access to the development platform used. Providing access without permission will be in breach of licence terms and terms of service.

Achieving the gold standard is thus potentially costly and difficult, and regulators might reasonably only demand it for the highest risk AI applications.[571]

Regulators might also take into account that consistency in the form of repeatability, replicability and reproducibility has close links to explainability. Explainability methods[572] can identify the data elements and their derivations which influenced an AI outcome, such as counterfactual analysis, which can identify where changes in weighting result in different decisions. From these results, data representing similar, but not identical, cases can be developed and used to test for reproducibility.[573]

[570] Benjamin Haibe-Kains et al, 'Transparency and reproducibility in artificial intelligence' (2020) 586 *Nature* E14. See also Pineau et al (n 565) 7463.

[571] Haibe-Kains et al, ibid, suggest that independent verification of reproducibility is essential for AI applications in medicine which determine patient treatment, such as breast cancer screening. A regulator might take a more nuanced view, balancing the costs of independent verification against the likely benefits from implementing the AI, and taking into account the reassurances about reproducibility which the AI producer is able to offer.

[572] See section 9.4. See also Reed, Grieman and Early (n 70).

[573] My thanks are due to Dr Keri Grieman for this insight, and for alerting me to the wider importance of consistency.

Thus for most AI applications, it seems that we cannot demand more from AI producers than that they use reasonable efforts to assess and demonstrate the reproducibility of their AI. This is exactly the same conclusion that I argued in respect to fairness, which suggests that justified risk mitigation explanations might be equally useful here as a regulatory tool. AI producers face many choices – which development platform offers the most appropriate technical facilities for assessing reproducibility; which functionalities should they use, and how; how far is it possible for them to make available code and data for independent assessment; and so on. Although regulation cannot specify those choices in advance, it is well adapted to assessing retrospectively the adequacy of an AI producer's own explanation and justification of those choices.

9.3. Trustworthiness

Human society runs on trust.[574] On my journey to work I drive to the railway station, trusting that the mechanic who serviced my car left it in a safe condition. On the train, I trust the driver and the signalling staff, and those who design and maintain the trains and railway system, to get me there safely. I trust that the coffee I buy is not poisonous or otherwise dangerous to my health. Without that trust, I would never leave home.[575]

My trust in the technology which enables my life, such as my car, the train and the coffee machine, is ultimately trust in the humans who design, build, operate and maintain that technology. I am forced to trust that they have ensured that the technology will do what it is supposed to do. Thus, as we introduce AI into my life, I will need to trust its creators and users in exactly the same way.

Trust is a complex phenomenon,[576] but for the purposes of this discussion I will conceptualise it as a belief in the future, a belief that humans will do what they have undertaken to or ought to do, and a belief that AI technology will work as intended.

Belief-based trust in humans is complicated. There are two possible sources of that belief.[577] The first is based on close relationships – family, friends, work colleagues. This kind of trust might be described as emotionally based. We believe that the other will do what they ought to do because of the nature of the relationship, and because of our past dealings with them in the relationship. We

[574] David J Lewis and Andrew Weigert, 'Trust as a social reality' (1985) 63 *Social Forces* 967.

[575] Luhman asserts that the only alternatives to trust are 'chaos and paralysing fear' – Niklas Luhmann, *Trust and Power* (Hoboken NJ, Wiley, 1979) 4.

[576] For an overview of the concept of trust, see Scott Thiebes, Sebastian Lins and Ali Sunyaev, 'Trustworthy artificial intelligence' (2021) 31 *Electronic Markets* 447, 449–51.

[577] For detailed discussion of these trust concepts see Elias L Khalil (ed), *Trust* (Cheltenham, Edward Elgar, 2003).

are unlikely to find this kind of trust in an AI because in most cases there is no emotional relationship with it or with its producer.[578]

The second source of belief is based on the existence of some kind of external controls which generate confidence that others will behave in the ways they ought to behave.[579] For example, the parties to a contract make promises to each other which aim to engender trust, and the possibility of legal enforcement through the courts creates confidence that the other party will perform their promises (or more accurately, that either they will perform or the law will provide a remedy if they do not). Liability laws, such as the law of torts, work in the same way. An alternative external control method is to establish a system of regulation where an independent body monitors compliance with rules and can impose sanctions on those who default, such as we find in data protection law or regulation of the medical professions. Thus, if we visit a new dentist, we will be likely to trust that dentist to an extent because their training has been certified and their performance is overseen by a regulatory body.[580]

Of course, repeated dealings can lead to the kind of relationship which engenders the first kind of trust, so that long-term business partners will often rely on each other's promises even in the absence of a contract. The same will be true of an AI – if it appears to have produced its intended outputs over a period of time, humans will begin to trust that it will do so in future. However, emotional trust engendered in this way is fragile; a single failure is likely to destroy it and focus attention again on the external controls.

Collectively, these individual trust relationships come together to produce the wider social trust which makes living in a community possible.[581] The question which needs to be explored, therefore, is what controls over human producers of AI will be likely to engender the second kind of trust amongst members of society, and how they should be implemented by that society.

The second kind of belief, leading to trust in technology, has very different roots. The primary source of such belief is a person's perceptions about the

[578] The exception might be an AI-controlled service robot, where it is conceivable that a one-sided emotional connection could develop. There is already a body of evidence in favour of this – see eg Gail F Melson, Peter H Kahn Jr, Alan Beck and Batya Friedman, 'Robotic pets in human lives: Implications for the human–animal bond and for human relationships with personified technologies' (2009) 65 *Journal of Social Issues* 545; Markus Blut, Cheng Wang, Nancy V Wünderlich and Christian Brock, 'Understanding anthropomorphism in service provision: a meta-analysis of physical robots, chatbots, and other AI' (2021) 49 *Journal of the Academy of Marketing Science* 632.

[579] Lewis and Weigert (n 574) 973 describe this as the 'system trust which modem society increasingly depends on, inasmuch as most of the interactions occurring in such societies would be too risky, unpredictable, or downright impossible if they had to be based only on personal trust.'

[580] Repeated visits to the same dentist might generate some level of emotional trust, derived from previous good experiences, or might reduce the initial trust produced by regulation if the dentist's work is not satisfactory.

[581] Lewis and Weigert (n 574) 968: 'trust exists in a social system insofar as the members of that system act according to and are secure in the expected futures constituted by the presence of each other or their symbolic representations.'

technology itself, in terms of its functionality, helpfulness and reliability.[582] To take the familiar example of a spell-checker, a user might ask whether it does in fact detect mis-spellings, how helpful it is in suggesting corrections, and whether it consistently performs those functions. The user's perception of the answers will determine how much trust is placed in the spell-checker. Thus the belief which leads to trust derives, initially at least, from an assessment of the capabilities of the technology.

However, there is still an emotional element to trust in technology. McKnight et al point out that:

> Because technology lacks volition and moral agency, IT-related trust necessarily reflects beliefs about a technology's characteristics rather than its will or motives, because it has none. This does not mean trust in technology is devoid of emotion, however. Emotion arises whenever a person's plans or goals are interrupted. Because we depend on less than reliable technology for many tasks, technology can interrupt our plans and raise emotion. For this reason, trust in technology will often reflect positive/negative emotions people develop towards a technology.[583]

These emotions are shaped by individuals' existing emotional make-up. Some people are technophilic, tending to trust technologies, while others begin from a position of distrust. Attitudes to those who produce technology and attempt to ensure its functionality also vary in terms of the trust placed in them. And finally, an individual's relationship with a technology based on their experience of using it also generates an emotional response to the technology.[584]

It is worth noting that the idea of trust in technology runs counter to the findings of social scientists that most humans do not trust technology per se, but rather they trust it because they trust its designers and producers.[585] This might be because humans begin from an initial state of ignorance when faced with a new technology – they know nothing of its functionality, helpfulness and reliability, and can thus form no trust-forming beliefs about it until they learn more. Afroogh suggests that trust in AI is likely to be probabilistic, based on a belief that the AI will probably (though not certainly) act as it is supposed to, and that this belief can be reinforced or reduced by experience of how the AI actually performs.[586]

[582] D Harrison Mcknight, Michelle Carter, Jason Bennett Thatcher and Paul F Clay, 'Trust in a specific technology: An investigation of its components and measures' (2011) 2 ACM *Transactions on Management Information Systems* 1, 5.

[583] ibid, 5.

[584] For deeper analysis of these issues see ibid, 6–10.

[585] See Adriano Fabris, 'Can we trust machines? The role of trust in technological environments' in Adriano Fabris (ed), *Trust: A Philosophical Approach* (Cham, Springer, 2020) 123, 134: 'Trust is born – as a human attitude towards someone or something that feels reliable – only within a relationship. Trust is a specific quality of such relationship. It is the way such relationship materialises.'

For a similar understanding of trust in the, then, new technological activity of e-commerce see Jason Rutter, 'From the sociology of trust towards a sociology of "e-trust"' (2001) 2 *International Journal of New Product Development & Innovation Management* 371.

[586] Saleh Afroogh, 'A probabilistic theory of trust concerning artificial intelligence: Can intelligent robots trust humans?' (2023) 3 *AI and Ethics* 469.

9.3.1. Technical Controls

So far, trustworthiness has been seen by the AI research and development community as primarily an issue of trust in technology. The need for trust is accepted, and it is to be achieved by identifying its constituent elements in terms of functionality, helpfulness and reliability and then devising tests which measure whether an AI has successfully achieved those elements. An AI which passes all these trustworthiness tests is thus, by definition, trustworthy (at least by engineers).

Li et al[587] have surveyed the range of trustworthiness tests which the AI research community has developed, identifying that they concentrate on six main elements:

- Robustness. A trustworthy AI will produce acceptable outcomes even if there are some errors or missing or unexpected elements in the input data for that outcome. Robustness can be tested by manipulating input data and comparing the AI's outcome with that which it produces using the original data. A robust AI will not differ in unacceptable ways – for example, in a loan granting AI, a small change in the age of a loan applicant should not result in a large change in their suitability score. A robust AI will also be resistant to attempts to hack it, or to 'spoof'[588] it by manipulating the input data.
- Generalisation. AIs are, inevitably, trained only on a subset of all the possible data which relates to the domain of their decision-making. Generalisation is their ability to produce acceptable outcomes across that domain – an AI which claims to evaluate employment applications, but which in fact produces accurate evaluations only for some undefined subset of those applications, cannot be trusted. Evaluation of an AI's generalisability also helps define the boundaries of its domain of application – an AI used outside that domain can clearly not be trusted to produce acceptable outcomes.
- Explainability and transparency. If the reasons for an outcome can be explained, this assists in testing for the other elements of trustworthiness and also enhances the likelihood of social acceptance of the use of that AI. The ideal which researchers are seeking is 'a direct quantitative measurement of explainability' which unifies existing metrics,[589] though a quantitative measure of how far explanation of an AI will satisfy those outside the technical community seems unlikely to be achievable. Transparency provides information about the whole life cycle of the AI's development – design, training, testing and deployment – and both assists technical evaluation and enhances social acceptance.

[587] Bo Li, Peng Qi, Bo Liu, Shuai Di, Jingen Liu, Jiquan Pei, Jinfeng Yi and Bowen Zhou, 'Trustworthy AI: From principles to practices' (2023) 55 *ACM Computing Surveys* 1.

[588] An example might be attempting to deceive a face recognition system – Ramachandra Raghavendra and Christoph Busch, 'Presentation attack detection methods for face recognition systems: A comprehensive survey' (2017) 50 *ACM Computing Surveys* 1.

[589] Li et al (n 587) 8.

- Reproducibility. An AI which produces radically different outcomes for like cases can hardly be seen as trustworthy, and we have already seen in chapter 6 that such an AI would also be perceived as unfair. From a purely functional perspective, an AI which produces inconsistent decisions is not effective in carrying out the tasks for which is was developed, as discussed at section 9.2 above.
- Fairness. No-one trusts an unfair decision-maker, and the difficulties in understanding and achieving fairness have already been examined at length in this book.
- Privacy. An AI which makes decisions about individuals will need access to detailed information about them, much of which will be private and/or confidential. All decision-makers must preserve that privacy and confidentiality if they are to be trusted.

What is clear from this evaluation is that AI producers face choices about the ways in which they might attempt to achieve all these aspects of trustworthiness, and in how they might evaluate whether they have been achieved. At least at present, there are no definitive answers about which choices are the best. This is exactly the situation facing AI producers in relation to fairness, as we saw in chapter seven.

The desire of the AI research community for definitive answers is clear:

> Repeatable and quantitative measurements are the cornerstone of scientific and engineering progress. However, despite increasing research interest and efforts, the quantification of many aspects of AI trustworthiness remains elusive. Of the various aspects that we have discussed in this article, the explainability, transparency, and accountability of AI systems are still seldom evaluated quantitatively, which makes it difficult to accurately compare systems. Developing good methods of quantitative evaluation for these desiderata, we believe, will be an important first step in research on these aspects of AI trustworthiness as a scientific endeavor, rather than a purely philosophical one.[590]

The difficulty in finding such answers is because, as explained at the beginning of this section, trust arises from the relationships between the human individual giving the trust and the human AI producers who are responsible for the technology in which trust is sought. This suggests that any regulation aiming to achieve trustworthy AI should focus primarily on the behaviour of its producers, not the technology itself.

Trustworthiness also seems likely to change its nature with changing context, which again makes any quantitative measure difficult to achieve. For example, both my hairdresser and my surgeon apply sharp instruments to my body, but I need to trust the surgeon far more. We recognise this as a society by regulating surgeons heavily, while regulating hairdressers hardly at all. Context is a recognised obstacle

[590] ibid, 29.

to objective metrics for AI fairness,[591] and seems likely to present much the same difficulties for trustworthiness.

9.3.2. Human Controls and Accountability

The discussion above seems to rule out any idea of regulating AI producers by prescribing the technical steps they must take to achieve trustworthiness, except perhaps in those regulated sectors where standards can be developed. Even there, it is unlikely that trustworthiness can be achieved through technical measures alone.

There does appear to be a growing consensus that accountability on the part of AI producers plays an important part in achieving trust. AI substitutes for human activity, and is thus its trustworthiness is likely to be assessed more in terms of its human origins than its technical capabilities: 'research has shown that trust in a technology which is perceived as human-like by its user or is highly automated and autonomous, differs from classical interpersonal trust and also classical trust in technology'.[592] This means that trust in AI derives mainly from trust in those who produce it, which in turn depends on those producers creating a relationship with those from whom trust is needed.

This requires positive action on the part of AI producers. Avin et al propose that there is a need to create 'an ecosystem where AI developers can earn trust – if they are trustworthy',[593] and suggest that such trust can be earned via (inter alia) independent specialist review, external audits,[594] and explainability and transparency. These are all aspects of accountability: 'Accountability runs through the entire lifecycle of an AI system and requires that the stakeholders of an AI system be obligated to justify their design, implementation, and operation as aligned with human values.'[595]

Accountability of this kind is perhaps the most important way of ensuring that ethical considerations have been incorporated into the AI's production, and thus 'tracing moral responsibility and accountability for the actions performed by ML algorithms.'[596]

At present, the levels of this kind of justificatory accountability vary between AI producers, and thus the trustworthiness of their AI systems also varies. If consistency were desired by regulators, one way to achieve this would be to bring

[591] See section 7.6.
[592] Thiebes, Lin and Sunyaev (n 576) 450.
[593] Shahar Avin et al, 'Filling gaps in trustworthy development of AI' (2021) 374 *Science* 1327.
[594] A worked example of how a particular AI implementation, face recognition technology, might be audited is in Evani Radiya-Dixit and Gina Neff, 'A Sociotechnical Audit: Assessing Police Use of Facial Recognition' (2023) *Proceedings of the 2023 ACM Conference on Fairness, Accountability, and Transparency* 1334.
[595] Li et al (n 587) 12.
[596] Andreas Tsamados et al, 'The ethics of algorithms: key problems and solutions' (2021) *Ethics, Governance, and Policies in Artificial Intelligence* 215, 224.

the relevant elements of trustworthiness within the regulatory regime of risk mitigation explanations described in section 8.4.

This will be more complex than regulating fairness. Fairness is not easily decomposable into discrete elements, as we have seen, and so the regulatory obligation there is simply expressed as one of producing an explanation about identification of risks of unfairness, actions to mitigate them, and justification of those actions. By contrast, trustworthiness has distinct elements, as explained in section 9.3.1 and it may not be appropriate for regulation to demand that all of them be dealt with in a risk mitigation explanation. The context in which the AI is deployed will be relevant here.

Let us take as our example an AI which drives a vehicle. We clearly want to be able to trust that AI, which means trusting its producers, if it is to be let loose on the roads. From the list of trustworthiness factors in section 9.3.1 we are clearly very interested in its robustness, its ability to generalise to cope with new road conditions, its reproducibility in that it must drive consistently and predictably, and that we can discover and understand why it made its driving decisions in the event of any failure on its part. Fairness and privacy are almost certainly irrelevant matters, and the time and effort of both regulator and regulatee should not be wasted on them.

By contrast, the medical treatment allocation AI we have returned to throughout clearly requires all those trustworthiness factors to be considered if its decisions are to be trusted by medics and patients.

9.3.3. Generative AI and Untruths

A clear advantage of regulating via risk mitigation explanations is its flexibility when faced with new AI capabilities. One trustworthiness risk, which the literature in section 9.3.1 does not identify explicitly, is the risk that the AI produces deceptive outputs. The recent transformation of generative AI from a research activity to a service which is available for use by the general public has highlighted its potential to produce untrustworthy outputs, either because the AI user wishes to do so (as in the case of deepfake images and video) or because the AI technology only *simulates* truth, as does ChatGPT, for example, rather than seeking to produce truth.

Truthfulness is an aspirational value, in the case of generative AI, because there is no obvious route to achieving truthfulness which regulation might mandate. The best which regulation can ask for, in the current state of our knowledge about the technology and the ways in which it might be used, is that a generative AI producer should do the best they can to mitigate the risks that untruthful outputs damage the fundamental rights or economic interests of others.

Trust is an emotional response to a technology or its producers, as the discussion at the beginning of this section 9.3 explains. In the case of an 'ordinary' AI, that emotional response of trustworthiness is achieved by demonstrating that the AI is likely to exhibit other aspirational values like fairness, robustness, repeatability and accuracy. Unfortunately such demonstration is not feasible for generative AI, because of the known risks of hallucination,[597] the existence of untruthful sources in the training data, and the longer-term risk of 'model collapse' as more and more output of generative AI enters the training dataset.[598]

The risks of untruthfulness can be mitigated by providing warnings which are sufficient to reduce the likelihood of unwarranted reliance on the generative AI's output, or by AI users ensuring that they do not use generative AI for applications where untruthfulness might be problematic or implementing other protective measures such as fact-checking. The extent of these risks, and how they might be safeguarded against, is best identified and implemented by AI producers for their individual application – at best, an external regulator can only guess what generic risks might require action, and this is unlikely to identify the actual risks in a specific implementation or the best ways of dealing with them.

This raises the question of when and on whom risk mitigation obligations should be imposed. It is clearly not feasible to regulate the producers of foundation model generative AIs to require them to ensure that their outputs are truthful. Whether an output might deceive others, and whether this is a problem which society needs to address, depends entirely on the nature of the output and how it is used. For example, a deepfake image of a politician accepting a suitcase full of money is not problematic if I produce it for my own amusement, but it is a potential matter for regulation if I post it to social media claiming that it is an actual photograph of the politician. Medical advice from ChatGPT is clearly problematic if following it might damage a user's health, but the producers of ChatGPT have no way of knowing that this is what the user might do.[599]

All this suggests that we cannot realistically make regulatory demands that generative AIs tell the truth, but society might be justified in requiring some AI producers to identify the risks that their generative AI might mislead others and to take appropriate steps to mitigate those risks. In the same way that I have suggested that fairness regulation is only justified if unfairness creates serious risks to others, this obligation should only be imposed where the AI's untruthful output is likely to have serious consequences.

Under this approach to regulation, developers of foundation model generative AIs would have quite different obligations from AI users who adopt those foundation models, with or without modification, for some specific task. Foundation

[597] See n 557.

[598] This is where the output of a generative AI trained a dataset which includes generative AI output becomes increasingly inaccurate and tends towards nonsensical statements – see Shumailov et al (n 560).

[599] ChatGPT does implement some controls, designed to prevent it from providing output which is 'sexual, hateful, violent, or promotes self-harm' – ChatGPT Moderation blog post, openai.com/blog/new-and-improved-content-moderation-tooling. However, a web search for 'ChatGPT content moderation bypass' returns numerous pages explaining how to evade this moderation.

models have an unrestricted domain of use, so that they can, for example, generate text on any conceivable subject or produce images of any kind describable. Thus the foundation model developer will only be able to identify the generic risk that the AI produces misleading output, and will be unable to predict which outputs have potentially serious consequences because the use to which the AI will be put is determined by the AI user, not pre-assigned by the developer. This suggests that the developer cannot be expected to do more than to warn the user about the risk of misleading output, suggest ways in which it can be avoided, and exhort the user to undertake some independent check of truthfulness before relying on the output. If some user prompts or the corresponding output can be identified as having the potential to cause harm if misleading then the developer might also be expected to implement some controls over them to prevent that output being generated, as ChatGPT does.[600]

A downstream AI user who builds an application powered by generative AI will have more detailed knowledge about how the AI's output will be used, and thus a greater ability to predict whether that may produce serious consequences. If, for example, a generative text AI is used for customer support in relation to a product, the AI user can be expected to predict the ways in which that product might be misused as a consequence of the information supplied by the AI, and therefore to take additional steps to explain and mitigate those risks. Merely warning the customer that the AI's advice is not reliable will be insufficient, because the whole purpose of customer support is to be relied on and thus customers will be likely to ignore such warnings as merely formulaic.

If the generative AI is made accessible directly to the general public there is little more that regulation can do. Assuming that adequate warnings have been given by the developer, responsibility for any ill consequences of relying on outputs has to fall on the individual user. Generative AI is useful but dangerous in the wrong hands, as is electricity. In spite of its dangers we still make electricity available to untrained users, even though householders can easily bypass the regulatory safeguards put in place.

9.4. Explainability

Trustworthiness is closely linked to explainability. Users of technologies may not themselves be able to explain how the technology works, but a large element of their trust that it *does* work as intended is based on the assumption that the technology's makers are able to explain how it works.[601] Explanations about the workings of the

[600] ibid.

[601] The ability to explain not only helps engender trust in AI decisions, but also in the person who uses that AI for decision-making – see Stephan Grimmelikhuijsen, 'Explaining why the computer says no: Algorithmic transparency affects the perceived trustworthiness of automated decision-making' (2023) 83 *Public Administration Review* 241.

technology will be needed if it requires some regulatory approval, or in order to convince buyers that they should purchase it. Explanations of any failure to work correctly will be needed after the event, to decide what the consequences of that failure should be (for example, in a legal action for compensation, or a regulatory decision about whether the technology should be modified to maintain approval).

Most technologies might be described as deterministic, in that there is a limited range of possible inputs each of which always produces the same output. A car steering wheel, for example, can be turned only right or left, and the consequence is always that the wheels move right or left in proportion to that movement. If more detail is required, the mechanical or electronic linkages which achieve this output can also be explained. If the steering fails, it is possible to detect which component in the system led to that failure and thus explain *how* it occurred. Analysis of the component can explain *why* it occurred, perhaps through wear and tear or through defective design or manufacture.

AI developed through machine learning is problematic here. As we saw in section 7.1, even deterministic AIs are so complex that an explanation may be hard to construct. AIs based on neural nets use a learning process which detects patterns in the training data, and where such an AI is used, it bases decisions on how well the input data matches those patterns. This means that simple observation of the internal structure of the AI tells us almost nothing about how it will decide for a particular set of inputs. A small change to those inputs may change the pattern which is detected and lead to quite a different outcome. Thus although we can predict in advance what will, or at least should, happen if a car steering wheel is turned, we will find it far more difficult to make a similar prediction for many AIs. This means that any explanation in advance of how the AI will work, in terms of what the technology is doing internally, is at best likely to be incomplete and may also be very difficult for most audiences to understand.[602]

After the event, things are more hopeful. The fast-developing field of eXplainable AI (XAI) has the goal of designing tools that can provide retrospective explanations for the decisions of AI systems.[603] Each tool focuses on a different aspect of the AI's workings, and can broadly be categorised into three groups.[604]

[602] Insight into these difficulties can be found from the empirical investigation in Lauren Waardenburg, Marleen Huysman and Anastasia V Sergeeva, 'In the land of the blind, the one-eyed man is king: Knowledge brokerage in the age of learning algorithms' (2023) 33 *Organization Science* 59 into the implementation of predictive policing AI by a Dutch police force. The AI developers were unable or unwilling to explain why the AI made its recommendations, and this led police staff to reverse engineer their own explanations for its outputs from their broader knowledge of the area and its population in order to make them usable for policing strategy.

[603] For an introduction to XAI see Derek Doran, Sarah Schulz and Tarek R Besold, 'What does explainable AI really mean? A new conceptualization of perspectives' (2018) *CEUR Workshop Proceedings*, vol 2071; Sule Anjomshoae et al, 'Explainable agents and robots: Results from a systematic literature review' (2019) *18th International Conference on Autonomous Agents and Multiagent Systems (AAMAS 2019), Montreal, Canada, 13–17 May 2019*; Nadia Burkart and Marco F Huber, 'A survey on the explainability of supervised machine learning' (2021) 70 *Journal of Artificial Intelligence Research* 245.

[604] The terminology in the XAI field is still developing, and the labels used here are taken from Erico Tjoa and Cuntai Guan, 'A survey on explainable artificial intelligence (XAI): towards medical XAI'

The first group of tools attempts to evaluate the linkages between the inputs to an AI and its outputs, using a range of techniques. Saliency techniques identify the relative importance of different inputs to the AI in producing particular outputs. In our medical treatment allocation AI example, these tools might identify that the age of the patient played an important part, whilst the fact that they also suffered from another medical condition had little influence in the outcome. Results from these tools are often presented visually as graphs or heat maps. The idea here is that humans will be able to identify patterns in those results and then map them to their own understandings of how decisions in that field are made. Feature extraction (or importance) techniques acts in a similar way by taking features in the input data (eg for medical diagnosis, the inputs relating to fitness, eating patterns and sleep patterns) and then identifying across a range of decisions the features which are most strongly correlated for particular outputs and those which are not correlated. Feature extraction is thus a type of abstraction; it might find, for a particular disease, that when the AI makes its diagnoses sleep and diet are closely correlated, whereas geographical residence and income are not. These correlations can assist in devising human-understandable explanations about how the AI works. Sensitivity methods test the effects of inputs on outputs using a counterfactual approach, by taking individual decisions of the AI and making changes to its inputs to see how they affect its outputs.[605] This can identify which inputs are most important for producing the decision. It can also offer a measure of reliability for the AI, because if a tiny change in inputs produce major changes in output, the AI might not produce reliable results on inputs it has not seen before. A challenge with all these techniques is generalising them to provide useful information about the workings of the AI overall, rather than just explaining individual decisions.

The second group of tools attempts to produce interpretable models whose outputs appear to match closely those produced by the AI. These models might be rule-based if the nature of the AI makes that appropriate, for example AI text processing, or mathematically based if the AI works via pattern matching. One difficulty here is in explaining and justifying the choice of modelling technique, which can be equally as opaque as the AI which it attempts to explain. The aim of modelling is to abstract the human-incomprehensible workings of the AI into

(2020) 32 *IEEE transactions on neural networks and learning systems* 4793. For alternative taxonomies see eg Or Biran and Courtenay Cotton, 'Explanation and justification in machine learning: A survey' (2017) 8 *IJCAI-17 workshop on explainable AI (XAI)*; Riccardo Guidotti et al, 'A survey of methods for explaining black box models' (2018) 51.5 *ACM computing surveys (CSUR)* 1.

[605] Counterfactual explanations function by reiterating a data process with the smallest possible change to determine which parts of the data are influencing a decision. Small tweaks are made to the input data, then the 'question' put to the AI is asked again and again, pinpointing which data points changed the outcome. While an explanation tends to refer to 'an attempt to convey the internal state or logic of an algorithm … counterfactuals describe a dependency on the external facts that led to that decision.' Sandra Wachter, Brent Mittelstadt and Chris Russell, 'Counterfactual Explanations Without Opening the Black Box: Automated Decisions and the GDPR' (2017), www.ssrn.com/abstract=3063289. See also Atoosa Kasirzadeh and Andrew Smart, The Use and Misuse of Counterfactuals in Ethical Machine Learning (2021), arxiv.org/abs/2102.05085.

a simplified model which is understandable by some skilled humans, so that those humans can then explain the AI to other humans at an even higher level of abstraction. An alternative to modelling the entire AI is via optimisation (or decomposition), which attempts to find sub-elements of the AI which, for the same input data, produce outputs which are recognisably related to the full AI's output. This is also a kind of abstraction of the AI, and similarly assumes that the abstraction can be interpreted (probably by technical experts) to explain information about the full AI's decision-making.

The final group consists of signal methods, which are primarily used for image recognition neural nets. These tools identify how input images affect the values of the neurons in a layer of the network. What that layer 'sees' can then be reconstructed and compared by a human to the original image, to discover which parts of the input image are detected by each layer. Using this information, an expert might be able to construct a story about how the AI goes about identifying an image. In the case of, say, face recognition, the story might be: 'First the AI identifies the eyes and nose, the next layer finds the edges of the face, the third layer …'.

So how far do these tools explain how the AI has made its decision? The answer is: not very far on their own. Each of them appears to give us *some* information about the AI's decision processes, but it is clear that this information needs to be interpreted by an expert into the kind of explanation which the audience needs. We also need to recognise that none of the tools tell us with certainty what the AI has done – at best they tell us what it most likely did, and in some circumstances might be able to put a probability value on the tool's finding.

This then raises questions about the adequacy of the expert's explanation. An explanation has no abstract value – it is only useful if it provides the information which the person receiving it requires, which depends on who that person is and what assurances they need. A decision subject might need to know that the AI which made the decision was likely to have produced a fair outcome, that the decision was based on full and accurate information about the subject, and (at a high level) some justification for the outcome. A regulator would have different needs, would be interested in knowing how the AI was developed, trained and tested, and is more likely to focus on a collective set of outcomes than on individual decisions.

If we tried to analyse what humans tend to ask about technologies, we could summarise them in three broad questions:

- How will it work? This might be something a human needs to know before using or submitting to the technology. However, the question is not worded quite right. Technology is never infallible, and humans know that. A better way to phrase the question might be: how ought it to work? There is thus a normative element to this question, but also a justificatory element. By putting a technology into use and explaining what I intend it to do, I raise the expectation in those who will be affected that I have some grounds to believe that it will in fact work as I intend it to. So we can expand this question further into: how ought it to work; and why do you as its producer believe that it will do so?

- Why did it do what it did? We humans tend to leave the 'how did it do it?' question to technologists, and do not normally even ask the 'why' question unless we suspect that the technology should have done something different. Thus this question also requires an answer to the first, normative question. More accurately the question is: was it supposed to act in this way? We will generally ask this question to satisfy ourselves about the reliability, fairness, trustworthiness, etc of an AI, and as an answer we will expect to be told how it should have acted and, if it did not, some reason why.
- Why didn't it do what it should have done? This is the converse of the previous question and is clearly also dependent on the first question. It requires the same kind of answer.

Thus we are seeking two different types of explanation: a normative description of how the AI was intended to act; and, in the case of actual or suspected departure from that intended action, the reasons why the AI did so.

XAI tools offers us no help with answering the first, normative question. Instead, we must look to the AI developer and the AI user to tell us their intent. We also look to them to justify their belief that the AI works as intended, and this can be achieved by an explanation of the development, training and testing regime which was adopted, framed appropriately for its intended audience. Part of that justification might include the ways in which XAI tools were used to identify and remedy failures during training and testing, but the tools themselves will play no part in creating the answer. The final explanation of how the AI ought to work needs to be based on the risk analysis and mitigation measures taken by developer and user.

The second and third questions ask for an explanation why the AI failed to achieve its normative intent, either because it produced an outcome which it should not have produced, or failed to produce an outcome which it should have produced. When we seek an explanation from a human who is responsible for an outcome we expect answers in the form of a narrative, explaining how the human went about making the decision which led to that outcome. Then we can compare this answer to our chosen standard of human behaviour, such as taking reasonable care. This suggests that non-technologist audiences will want the same kind of narrative to explain an AI's actions.

Using XAI to construct such a narrative will be a different process from a narrative which explains how a human reached a decision. Human decision-makers review their thoughts, identify those which seem most salient, and then weave them into a story which creates causal links between each and leads to the outcome. XAI tools offer probabilities, not clear causal assertions, so that the resulting explanation will be about the most likely factors which influenced the outcome, rather than a clear narrative of causality. Thus XAI might identify which inputs most strongly influenced the final decision and which had little effect, or even the order in which an AI built up its decision based on its inputs, but is unlikely to be able generate an explanation about the AI's 'reasoning' because the

AI is not deciding in the same way that a human would. A human interpreter can take the outputs of XAI tools and weave them together to create something which approximates to the narrative which a human would give, but this is not the AI explaining itself – it is a human doing so, generating an human-like explanation of the AI's decision, based on observation of its workings by XAI tools.

Providing an explanation in advance about how the AI will act is even more challenging. The obvious starting point for technologists is to offer metrics about the AI's performance overall, but these metrics alone do not provide satisfactory explanation for other audiences, such as regulators and the general public. For example, the obvious question which autonomous vehicles pose is whether they are adequately safe. Some metrics are already available from California, where extensive on-road testing has been permitted. An analysis by Wang et al[606] of the 2014–19 data generates the following metrics:

- autonomous vehicles were involved in 4.9×10^5 accidents per mile, or 4.9 per 100,000 miles.
- 93.7 per cent of these accidents were caused not by the autonomous vehicle but by the actions of other road users.

From this we can calculate that the number of accidents caused by autonomous vehicles was 0.3 per 100,000 miles. Figures for human drivers can be calculated for the entire US population from the US Department of Transport statistics, which show that in 2019, humans were involved in 0.2 reported accidents per 100,000 miles.[607] It seems likely that all autonomous vehicle accidents were reported because of regulatory requirements, whereas some human accidents will not have been reported, so it is probable that in 2019 autonomous vehicle accident causation was broadly comparable to that of human drivers. Since then autonomous vehicle driving is likely to have improved, whereas human driving is likely to be unchanged in quality, so by now autonomous vehicles may be slightly less likely to cause accidents. Additionally, around a quarter of human-caused collisions resulted in injury or fatality,[608] a figure which is not available for autonomous vehicles but seems likely to be far lower.

These numbers suggest that autonomous vehicle driving generates at least no more risk than human driving, but cannot answer whether it is adequately safe or when it would be so. Safety, like fairness and trust, is not an absolute value but an emotional response to perceived risk.[609] Metrics alone are unlikely to generate that

[606] Jun Wang, Li Zhang, Yanjun Huang, Jian Zhao and Francesco Bella, 'Safety of autonomous vehicles' (2020) *Journal of Advanced Transportation* 1, 8.

[607] Calculated by the author from figures in crashstats.nhtsa.dot.gov/Api/Public/Publication/813060.

[608] ibid.

[609] See eg Robyn S Wilson, Adam Zwickle and Hugh Walpole, 'Developing a broadly applicable measure of risk perception' (2019) 39 *Risk Analysis* 777; Ellen M Peters, Burt Burraston and CK Mertz, 'An emotion-based model of risk perception and stigma susceptibility: Cognitive appraisals of emotion, affective reactivity, worldviews, and risk perceptions in the generation of technological stigma' (2004) 5 *Risk Analysis* 1349.

emotional response, particularly where a technology, such as AI, generates unpredictable risks. For example, an important cause of the well-known Tesla crash in the US in 2016 was that the autonomous driving technology misidentified another vehicle as being part of the sky, and so did not brake or turn to avoid collision.[610] Society expects that no human driver ought to make such a mistake, or rather, make such a mistake for this reason.

Thus advance explanations of how an AI is likely to act will need to be in narrative form. They will begin by explaining how its producers intended the AI to act, based on design, training and testing. Performance metrics will need to be combined with information from XAI tools, and the result will be an abstracted, high-level explanation of how that particular AI is likely to act. We will need to recognise, though, that this is only a model of what the AI might be doing, developed from sample cases and the result of human interpretation of an XAI analysis of the AI's workings. The model might work only for some types of case, and not for others, so this will be a dynamic explanation – over time the model might be disproved and information from metrics and XAI used to identify some other model, or that information might be used to improve the model so that its explanation is reasonably accurate for more cases. Such a model is only the best guess that can currently be achieved about the AI's 'reasoning', a mixed product of machine analysis and human interpretation.

How, then, does all this translate into regulation? We already know that there will be regulatory demands for AI transparency, as for example in the EU AI Act,[611] Article 13 of which provides that 'High-risk AI systems shall be designed and developed in such a way to ensure that their operation is sufficiently transparent to enable developers to interpret the system's output and use it appropriately' and also requires sufficient transparency to permit providers and deployers[612] to meet their own regulatory obligations.

Meaningful transparency can only be given in the form of an explanation, and the analysis above suggests that there are two important elements to such an explanation:

- how the AI was intended by its producers to operate; and
- the risks that it might fail to operate as intended and the reasons for failure if it does so.

To explain the second element requires identification of risks. Each AI generates its own particular risks, which suggests that it is infeasible for a regulator to define in advance the risks which must be explained. Instead, regulators will need to rely on risk analysis by the AI's producers.

[610] Larry Greenemeier, 'Driverless Cars Will Face Moral Dilemmas', *Scientific American* (23 June 2016), www.scientificamerican.com/article/driverless-cars-will-face-moral-dilemmas; Tesla Motors statement (30 June 2016), www.teslamotors.com/en_GB/blog/tragic-loss.

[611] EU AI Act (n 466).

[612] Providers are those who make AI systems available on the market, and deployers are the equivalent of AI users in this book – ibid, Art 3.

Explaining risks alone will not be enough. Those who are AI users, or who are subjects of an AI's decisions, also need to know what steps have been taken to mitigate those risks. Without such an explanation, it is not possible to know how far the AI can be relied on to produce its intended output, and thus whether it can be trusted and is likely to be fair. Again, AI producers have a choice of tools to help them explain, and ways in which they can use the outputs of those tools in their explanations, and are better placed to make the right choices than any regulator could be.

The meta-regulatory approach proposed in chapter eight for AI fairness is based on explanation of risk-mitigation, and thus incorporates the issue of explainability. This would also be true if risk-mitigation explanations are also required for other matters, such as trustworthiness. In all cases the regulatory obligation will be to explain, which can only be done if the AI is sufficiently explainable. There is no need to regulate to require explainability in any particular form, because an AI which is not explainable will prevent its producer from producing an explanation which satisfies the regulation. The choice of explainability tools and techniques can thus safely be left to the AI producer, because that choice is one of the matters which will need to be included as part of the justification for risk mitigation measures.

9.5. Use of Intellectual Creations for AI Development

Generative AI applications which produce text, images, video and music in response to user instructions have created real concern among those whose intellectual creations are used to train those AIs. Their main worries are that:

- use of their creations for training might infringe the intellectual property rights in those creations, primarily copyright and image rights (in countries where image rights exist);
- their creations might be, in part, reproduced in the AI's output; and
- the AI's output might substitute for their own creations, thus reducing the value of their existing creations and the demand for future creations.

None of these worries fall clearly within the scope of ethics, but they clearly indicate the existence of a social conflict of the kind that law and regulation aim to resolve.

As just two examples, consider book publishing and movie-making. A publisher might foresee a market for a sequel to a novel by a deceased author, in which case having it written by an AI will be cheaper and quicker than by a human author. If the original novel is still in copyright, the copyright owners will as a minimum want some share in the proceeds, and might well want some control over the new work in order to protect the literary reputation of the original author. In movie-making, one of the biggest costs is the fees paid to movie stars and the investment of money, time and effort required to film them, so a new movie starring AI-created versions of Humphrey Bogart or Marilyn Monroe might prove both successful and profitable. Image rights and trade mark law would be the only obstacles. And of course there is no reason, other than infringement of legal rights, why the original creators or performers need to be dead for such AI works to be created.

Existing intellectual property laws apply to such activities, but not in any coherent way. This is unsurprising – when those laws were devised, generative AI was unforeseen. Over time litigation will be able to establish how existing laws apply, but it is already clear from the extensive analysis by scholars of these issues[613] that generative AI challenges the fundamental conceptions underlying the law, particularly copyright, to such an extent that it is unlikely to be capable of solving these problems.

9.5.1. Copyright

The underlying conception of copyright law is that it should protect the intellectual creations of authors (including painters, photographers, composers and so on).[614] However, defining what amounts to an intellectual creation is too difficult a task for legislative drafting, and so what the law protects is a proxy for creativity, the *expression* of those creations in a work. It does this by granting the author exclusive rights to control copying and the exploitation of copies.[615] The underlying *idea* of the work is not part of its expression, and so is not protected – this is fundamental to the copyright bargain with authors, that in return for their exclusive rights others must be free to learn from, be inspired by and build on their works.

Using the works of others to train a generative AI often falls within the existing conception of copyright law. This is because the process of data ingestion, in which the work is checked and cleaned up if necessary and then incorporated into the training dataset, requires a copy to be made. Copyright law therefore asks two questions to decide whether this copying infringes.

[613] See eg Carys J Craig, 'The AI-copyright challenge: Tech-neutrality, authorship, and the public interest' in Ryan Abbott (ed), *Research Handbook on Intellectual Property and Artificial Intelligence* (Cheltenham, Edward Elgar, 2022) ch 7; Giorgio Franceschelli, and Mirco Musolesi, 'Copyright in generative deep learning' (2022) 4 *Data & Policy* e17; Jessica L Gillotte, 'Copyright Infringement in AI-Generated Artworks' (2020) 53 *UC Davis L Rev* 2655; Michael D Murray, 'Generative AI Art: Copyright Infringement and Fair Use' (2023), papers.ssrn.com/sol3/papers.cfm?abstract_id=4483539; Matthew Sag, 'Copyright safety for generative AI' (2023), papers.ssrn.com/sol3/papers.cfm?abstractid=4438593; Eric Sunray, 'Sounds of Science: Copyright Infringement in AI Music Generator Outputs' (2021) 29 *Cath U J L & Tech* 185.

[614] *Infopaq International v Danske Dagblades Forening*, Case C 2009:465 (EU); *Feist Publications Inc v Rural Telephone Service Co*, 499 US 340 (1991).

[615] When the internet technologies arrived in the mid-1990s these, too, challenged copyright law, and so an additional right to control communication of a work to the public (by 'wire or wireless means') was added by Art 8 of the WIPO Copyright Treaty. However, this merely identified a new way to exploit a work rather than changing the underlying conception.

First, is the copying unauthorised? It might not be if the copyright owner made it available under licence terms which permit the copying, which appears to be what happened in the US litigation between unnamed software developers and GitHub and OpenAI, in progress at the time of writing.[616] GitHub hosts open-source software code, and OpenAI used the GitHub code database to train its Codex generative AI. Codex writes computer code in response to user prompts about what the code needs to achieve. The complaint does not allege that OpenAI made infringing copies of the claimants' code, and this is likely to be because the GitHub terms of service, at least arguably, allow such copying. As explained by Samuelson:[617]

> GitHub requires all of its users to agree to its terms of service. Included in these terms is a license granting GitHub the right to 'store, archive, parse, and display … and make incidental copies' of the users' code, as well as to 'parse it into a search index or otherwise analyze it' and 'share' the resulting code in public repositories with other users.
>
> And when GitHub users decide to make their code available on its site, they must choose whether to make their code repositories private or public. Users who decide to make their repositories public 'grant each User of GitHub a nonexclusive, worldwide license to use, display, and perform Your Content through the GitHub Service and to reproduce Your Content solely on GitHub as permitted through GitHub's functionality.'

Many works are made available via third-party platforms, such as YouTube or Instagram, and these invariably give the platform owner rights to make copies. In each case the wording of the licence terms will define how far the platform owner can partner with an AI developer and allow copying of the works for training.

If the copyright owner simply makes the work available via its own website there will be no such express licence terms. By doing this the owner grants to the whole world an implied licence to view, play or listen to the work, but probably not a licence to copy the work and incorporate it into a training dataset.[618] However, if the generative AI could be trained on the work without downloading a copy, that might fall within the terms of the implied licence and thus not be an infringement.[619]

The second question is whether the copying falls within one of the exceptions to the copyright owner's exclusive rights. Probably the best known is the fair use

[616] dockets.justia.com/docket/california/candce/4:2022cv06823/403220.

[617] Pamela Samuelson, 'Legal Challenges to Generative AI, Part I' (2023) 66 *Communications of the ACM* 20, 23.

[618] I write 'probably' because the terms of such an implied licence can only be decided by a court, and then only for the particular case before the court. The Australian case of *Trumpet Software Pty Ltd v OzEmail Pty Ltd* [1996] 34 IPR 481 illustrates how a court might take into account the context of the making available (in this case via a disk distributed with a magazine), and any restrictions the copyright owner attempts to impose, when deciding the extent of an implied licence.

[619] However, website owners can give notice to AI developers revoking that licence, and also use their technology to prevent access, as has been reported for news organisations including the BBC – The Verge, 'The BBC is blocking OpenAI data scraping but is open to AI-powered journalism' (6 October 2023), www.theverge.com/2023/10/6/23906645/bbc-generative-ai-news-openai.

exception under section 106 of the US Copyright Act 1976;[620] other states have similar, though more restricted, exceptions. Samuelson suggests that if unauthorised copying had been alleged in the GitHub litigation, the defence of fair use might well have been successful. She analogises the case to *Author's Guild v Google*[621] where Google scanned (and thus copied) millions of books, many of which were in copyright, in order to build its Google Books service, which provides a searchable index and returns relevant snippets of the books. This copying was held to be fair use because it transformed the original books into something new and different which did not unduly substitute for or compete with the originals.

The fair use argument might be successful where a generative AI developer scrapes content from the internet and incorporates it into a training dataset. Although this content scraping is clearly copying, the resulting AI contains none of the expression of that content, and is certainly new and different from the content. Admittedly the output of the AI might sometimes come close enough to the original content that it substitutes for or competes with it, but the AI itself does not.

If an output of the AI, in part, appears to incorporate some of an original work which was used in training, the question whether this falls within the fundamental conception of copyright law is more difficult. Once a generative AI has been trained, all that it then 'knows' are the statistical relationships between the elements which it uses to build its output – words in the case of a text-generating AI, pixels for an image-generating AI, etc. In response to user prompts, the AI builds an output using these statistical relationships.

If we return to the example of a sequel to a novel, written by a new human author, we can see how the copyright conception applies to the creative process. The new author will have to read the existing novel, purchasing a legitimate copy to do so. The process of reading creates no copy, and nor does the process of familiarisation with the original author's plotting and writing style. Copyright law only becomes engaged as the sequel is written, and asks whether the new author has taken enough of the expression of the original to have infringed, or whether the new author has merely taken the idea which underlies the expression. Deciding this question is complex[622] – individual characters tend not to receive copyright protection per se,[623] but incorporating too much of the cast of the original plus some of its plot, as background to the new story, might well be enough to amount to copying a protected element of the original work. However, it will always be certain that if these elements of the original work appear in the sequel, there was a conscious decision by the new author to include them. They were undoubtedly

[620] Mark A Lemley and Bryan Casey, 'Fair learning' (2020) 88 *Texas Law Review* 743; Benjamin LW Sobel, 'Artificial intelligence's fair use crisis' (2017) 41 *Columbia Journal of Law and the Arts* 45.

[621] 721 F 3d 132 (2nd Cir 2015).

[622] For a detailed analysis under US copyright law see David B Feldman, 'Finding a home for fictional characters: A proposal for change in copyright protection' (1990) 78 *California Law Review* 687, 691–704.

[623] *Warner Bros Pictures v Columbia Broadcasting System*, 216 F 2d 945 (9th Cir 1954), cert denied, 348 US 971 (1955).

copied, and the question is whether that copying takes sufficient expression from the original to infringe.

Now imagine a similar sequel, written by a generative AI. The user prompt will have told it to mimic the style of the original author, which characters must be reused, and some indication how the plot should build on the original novel. These might be seen as copied from the original, but are probably not part of its expression. The remainder of the AI's output will be based on the statistical relationships between words – given the prompt instructions, what word is most likely to follow the current word?[624]

This looks nothing like copying the original expression. It is much more akin to independent creation, and courts around the world have consistently held that independent creation of the same or similar expression is not copyright infringement.[625] As Guadamuz puts it:

> [W]hen asking an AI tool to generate a work, it does not go back to the training data, it is working from a leaner and much more concise set of abstract concepts. But the result is the same, this is not a copy of a work, and it is not a derivative of a specific work.[626]

The uncertainty here is that the apparently independent creation has been influenced by the expression of the original, but also by pastiches and parodies of the original by other authors and, to a lesser extent, by the expression of every other author's work included in the training dataset. Is this influence enough to cross the line between independent creation and copying? Or is it more akin to one author taking inspiration from another? These are questions which can only be answered by anthropomorphising the AI, analogising its processes to those of a human mind. Copyright's fundamental conception that what is protected is intellectual creativity fails when presented with an AI which is neither intellectual, nor creative, in the human sense.[627] We would struggle just as hard to decide whether a recording of a parrot singing a song it has learnt from the radio[628] was a copyright infringement – copyright law was simply not designed to answer that question.

[624] For an explanation of the generative AI training process and its outcome see Andrés Guadamuz, 'A Scanner Darkly: Copyright Infringement in Artificial Intelligence Inputs and Outputs' (2023) 7–12, papers.ssrn.com/sol3/papers.cfm?abstractid=4371204.

[625] See eg US judge Learned Hand: 'the law imposes no prohibition upon those who, without copying, independently arrive at the precise combination of words or notes which have been copyrighted.' *Fred Fisher Inc v Dillingham*, 298 F 145, 147 (SDNY1924); for an equivalent UK decision see *Fay Evans v John Lewis PLC and another* [2023] EWHC 766 (IPEC).

[626] Guadamuz (n 623) 12.

[627] The US Copyright Office has determined that the law only grants copyright protection to works of human authorship, and has refused copyright registration for works which do not disclaim ownership of AI-created elements such as images. See *Review Board Decision on A Recent Entrance to Paradise* (14 February 2022), www.copyright.gov/rulings-filings/review-board/docs/a-recent-entrance-to-paradise.pdf; *Registration Decision on Zarya of the Dawn* (12 February 2023), www.copyright.gov/docs/zarya-of-the-dawn.pdf; *Review Board Decision on Théâtre D'opéra Spatial* (5 September 2023), www.copyright.gov/rulings-filings/review-board/docs/Theatre-Dopera-Spatial.pdf.

[628] See eg *Parrot singing If I was a boy by Beyoncé*, www.youtube.com/shorts/ik3mngsIC3c.

9.5.2. Unfair Competition

The third issue of concern – that an AI's output which is sufficiently similar to that of an existing author might reduce the demand for, and thus the value of, their future creations – is not really a copyright question. That output does not involve copying any of the expression of the works used in training. Copyright might be relevant if the output could be seen as a 'derivative work', such as a play or movie screenplay based on a novel, but for this to be arguable as an infringement some of the expression of the original (such as the plot of the novel) needs to be copied.[629]

Where no copying has taken place, this issue of reduced demand might be addressed by the unfair competition laws of some countries. Most states have some form of unfair competition law which provides a remedy where a rival business appropriates the trade reputation of another, such as by imitating the appearance of products or service provision,[630] following Article 10bis of the Paris Convention.[631] This is of little assistance to those whose works are used for training generative AIs, because the AI's outputs do not imitate the training materials. However, a minority of states extend their unfair competition laws to prohibit wider parasitic appropriation, which might best be understood as the misappropriation of another's intellectual work product[632] for business advantage.

Belgium is one such state, and in *Google v Copiepresse*[633] a Belgian court held that Google News, which displayed links to news stories on other sites together with the first few lines of the story, was in breach of unfair competition law[634] because Google was using the work product of the news publishers to create profits through its accompanying advertising sales. Some (though not all) US States take the same position, for example New York State: 'The principle that one may not misappropriate the results of the skill, expenditures and labors of a competitor has … often been implemented in our courts.'[635]

The difficulty in applying these principles of parasitic competition to generative AI arise because of the fundamental principle of copyright law, which is that in return for protection of the expression of a work, the ideas set out in that work should be free for all to use. This issue was raised before the US Supreme Court in *International News Service v Associated Press*,[636] where the defendant rewrote news stories gathered by the claimant and published them in its own news service.

[629] See eg *Tinsley v Lacy* (1863) 71 Eng Rep 327 (KB).

[630] See eg the New York state case of *ITC Ltd v Punchgini Inc*, 9 NY 3d 467, 476 (2007), where a New York restaurant imitated the name, menu, décor and staff clothing of the famous Bukhara chain of restaurants.

[631] Paris Convention for the Protection of Industrial Property (20 March 1883).

[632] Use of physical work product, such as land or chattels belonging to the claimant, would ground an action in trespass or the torts relating to interference with goods.

[633] Brussels Court of First Instance (TGI), 13 February 2007.

[634] And also in breach of copyright law, which is the main issue discussed in most analyses of this case.

[635] *Electrolux Corp v Val-Worth Inc*, 6 NY 2d 556, 567 (1959).

[636] *International News Service v Associated Press*, 248 US 215 (1918).

The Court held that news stories are facts, and thus unprotected by copyright; if they had been copyright-protected, state unfair competition law could not have applied. But because what was taken was not subject to copyright, and it amounted to 'material that has been acquired by complainant as the result of organization and the expenditure of labor, skill, and money, and which is salable by complainant for money',[637] appropriation and resale by the defendant was a breach of unfair competition law.

Thus to make a parasitic competition claim, the owner of a work which was used to train a generative AI would need to identify what the training process had taken from the work and reproduced in its output, that what was taken amounted to some form of business asset,[638] and probably also that what was taken would not more appropriately be protected by copyright. Additionally, some courts have limited the scope of protection to appropriation which directly competes with the claimant, and where allowing appropriation would disincentivise the claimant from investing effort in generating the asset.[639] Overcoming these obstacles to a claim would clearly be very difficult for most owners of works used in training.

9.5.3. Regulatory Solutions

At present, neither copyright law nor unfair competition law provide adequate solutions to the concerns which generative AI raises for intellectual creators. Society might, of course, conceivably take the view that no action is required. From one perspective, generative AI is doing the same things that competing authors do – analysing previous works, identifying the techniques for composition and expression which produce successful works, and then putting them into practice to create new works. The law has not previously interfered with those activities, and society is often argued to benefit from the widest possible level of production of works.

However, the level of concern which generative AI has produced is at a far higher level than previous concerns about human competition. Intellectual creators, at least at the functional level of instruction manuals, film and TV scenarios, graphic design and so on, fear that their future activities might be industrialised just as were those of hand-weaving and spinning,[640] limiting the market for human intellectual creation to what might be described as the 'luxury' end. Those who have already produced commercially exploitable creations fear that generative

[637] ibid, 239.
[638] In *International News Service* the court described these news stories as a form of 'quasi property', ibid 242.
[639] See *National Basketball Assn v Motorola Inc*, 105 F3d 841, 845 (2d Cir 1997).
[640] See Steven E Jones, *Against technology: From the Luddites to neo-Luddism* (New York, Routledge, 2013).

AI is a new way to 'free ride' on their creations without paying. Society might well decide that, at least during the transition to whatever the creative sphere will become, some limitations on generative AI are necessary to reduce the social disruption.

One possibility might to be to reform copyright and unfair competition law to deal with these issues, but in my view this is highly unlikely to be successful. As we saw, copyright law explicitly excludes ideas from protection, limiting protection solely to expression. What creators are primarily concerned about is the use by AI of their ideas, not their expression. Change would also be slow, as copyright law is notoriously difficult to reform because of its basis in international conventions.[641] The existence of these conventions severely limits the scope for national law reform, and securing international agreement to changes takes decades.[642] Additionally, such agreements are likely to be limited to extending the reach of copyright law without introducing new exceptions and limitations. Unfair competition law is also conceptually difficult to reform – extending its scope to 'parasitic' uses of works which do not compete directly with those works might introduce new and unforeseen problems in other areas.

The alternative is to produce regulation which deals specifically with these problems, so that its scope is limited to generative AI and creative works. That regulation needs to be compatible with existing copyright and unfair competition law, because those laws are so difficult to change.

Meta-regulation in the form of risk mitigation explanations might be a suitable vehicle, at least in the short term, to deal with some of these issues. The first two concerns of intellectual creators – that their creations are used to train generative AI without recompense, and that unprotected elements of their creations are reproduced in the AI's outputs – are at heart complaints that those creators are not being treated fairly.[643] If society were to agree that it is unfair to train an AI without compensating those whose works are used, and to reproduce the most creative ideas of others in generative AI outputs, then these could be added to the list of risks for which AI producers are required to produce risk mitigation explanations. We have seen throughout this book that fairness cannot be broken down into rules, but rather requires an assessment of the whole context in which a decision (here, to use a work for AI training without payment) is made, and so extensive guidelines on fair AI training will be necessary to assist AI producers in meeting their regulatory obligations.

[641] Berne Convention for the Protection of Literary and Artistic Works (1971 Paris text), adopted 9 September 1886, S Treaty Doc No 99-27, 1161 UNTS 3; WIPO Copyright Treaty, 20 December 1996, WIPO Doc CRNR/DC/94.

[642] The WIPO Copyright Treaty 1996 has its origins in the 1989 Berne Protocol discussions (US Congress, Expert Report 105-25 – WIPO Copyright Treaty (WCT) (1996) and WIPO Performances and Phonograms Treaty (WPPT) (1996), 2) and was the first major update to international copyright law since the 1967 Stockholm agreement led to the Paris Act 1971 updating the Berne Convention.

[643] Unfairness was also the complaint of the original Luddites, protesting against new weaving machines which reduced their wages – Jones (n 639) 3, 47.

Society might well make such a decision. There are already suggestions that society is beginning to accept that use of another's intellectual work product is something which normally requires compensation. We see this in the disputes which arise when search engines aggregate third-party news stories to provide their own news service. Copyright and unfair competition litigation in Belgium led to an agreement between Google and Copiepresse to licence news content,[644] and a similar dispute in Australia was also resolved via licensing.[645]

However, it is worth noting that in these disputes: (a) there was evidence that consumers were using online news aggregation as a substitute for the original news sources; and (b) that in each country there was a representative news organisation which could negotiate licence terms. Generative AI is trained on a far wider range of content, including fiction, non-fiction, functional works such as reports and manuals, and so on. National legislators would need to decide who deserves payment for use of their content, and might well be influenced by how far use by generative AI producers reduces the market for the original works. Novelists and visual artists might receive little protection because generative AI output seems unlikely to reduce their sales, whilst historians and instruction manual writers might conceivably lose income and thus deserve some recompense. Representative bodies would need to be established to negotiate licences for those kinds of work which deserve recompense for their use in AI training. It is hard to predict how a national legislator might define which uses of creative content are unfair (in the social, not the copyright, sense). However, if legislators *were* able to do so then a duty to explain and justify how the AI producer has mitigated the risk of unfair use would clearly be a useful and flexible regulatory tool.

The third concern of intellectual creators – that in future their potential customers may use generative AI to produce creations rather than engaging the services of human creators – is less obviously one which society might want to address through regulation. Technological change has, throughout human history, substituted for human work. Intellectual creation, or at least some aspects of it, may be the next casualty. '[As] many dominant producers have discovered in industries ranging from buggy whips to mainframe computers, even an established monopoly is vulnerable to entry through innovation'.[646] As a general principle, it is not the

[644] Copiepresse Press Release, Belgian French-language news publishers, authors societies and Google reach partnership agreement (13 December 2012), www.copiepresse.be/images/file/Google/20121212 CopiepressepressreleaseEN.pdf.

[645] There resolution was forced by legislation, the Treasury Laws Amendment (News Media and Digital Platforms Mandatory Bargaining Code) Act 2021, which required designated digital companies to agree terms for their use of online news. Although criticised as favouring the largest news publishers, it appears also to have led to licensing agreements with smaller publishers – William Turvill, 'Support for Australia's News Media Bargaining Code grows as 24 publishing minnows secure Google deals' *Press Gazette* (1 June 2022).

[646] Richard J Gilbert and Steven C Sunshine, 'Incorporating Dynamic Efficiency Concerns in Merger Analysis: The Use of Innovation Markets' (1995) 63 *Antitrust Law Journal* 569, 578.

role of law and regulation to preserve employment by prohibiting the use of new technology. Law and regulation will of course, have a role to play in resolving any resulting social problems,[647] but until it becomes clear what those problems are any attempt to regulate is likely to aim at the wrong target.

9.6. The Future Shape of AI Regulation

Regulating AI is more challenging than perhaps for any other new technology. AI has the potential to be used across all aspects of human life, and the ways in which it will impact individuals and society are still unknown. Attempting to devise a comprehensive system of AI regulation at this stage is a recipe for abject failure.

Unfortunately, waiting to discover the ways in which AI produces adverse effects which need to be controlled through regulation is also a recipe for failure. AI systems are being deployed rapidly across the board, and have the potential to become so deeply embedded in human life that serious damage might be caused before their defects can be remedied. Some risks of AI are already known, some are foreseeable as highly likely to happen, and others are speculative but of real concern if they do arise.

Human society therefore needs to take a risk-management approach to AI regulation, adopting regulatory tools which are sufficiently flexible that they can cope with changes in risk as they develop. However, we will need a different approach to risk management from that which is normally adopted.

The problem is that each sector of human activity has its own individual approach to risk management. The activities carried out within that sector are definable, and for each it is possible to identify risks which ought to be managed. Regulatory obligations can then be imposed in an attempt to ensure that those risks do not eventuate, or at least that their likelihood is reduced to an acceptable level.

Each AI instance has the potential to be used across many, and in some cases, all sectors. This means that the kind of detailed activity/risk mapping currently conducted at sectoral level is an impossibly large task. It can, and will, be done in already regulated sectors such as medicine and finance, and new sectors requiring regulation will doubtless be identified as more AI systems are implemented. But at the wider level of society as a whole, we cannot expect to produce a meaningful list of risky AI-based activities and identify ways in which we can regulate to control those risks, as we already saw in section 8.6.

[647] If social problems do in fact arise. There is an arguable case that generative AI creations provide something new and additional to human creativity, rather than substituting for it, as was the case for photography which did not (as feared) substitute for paintings and drawings. See Kevin Kelly, 'Picture Limitless Creativity at Your Fingertips' *Wired* (17 November 2022).

Take, for example, the OECD AI Principles,[648] an internationally agreed statement of the issues which government and regulators will need to address. These set out five 'values principles':

- Inclusive growth, sustainable development and well-being;
- Human-centred values and fairness;
- Transparency and explainability;
- Robustness, security and safety; and
- Accountability.

If we conceptualise these as risks which regulation ought to control, we can see that they are so broad that no useful regulatory solutions could be devised in terms of specifying how AI producers should behave.

We could, though, reconceptualise the AI regulatory project in terms of outcomes. The OECD AI Principles are much less obviously risks to be avoided than they are a socially desired set of outcomes which we hope to see when AI systems are implemented. The regulatory aim should therefore be to secure these outcomes, which at first sight seems an equally impossible task. But if we reduce our ambition, for the present, to regulating to reduce the risk that they are not achieved, the project becomes far more feasible.

Broadly, there are three categories of outcome which human society will want AI to achieve. The first is that when AI is adopted as part of activities which are already recognised to be risky, the AI should not increase the risks of the activity and, ideally, should reduce them. We know which activities these are because they are already regulated at the sectoral level – motor vehicles, medicine and financial services, to give just three examples. We also know how those risks will manifest themselves: through vehicle crashes, or patient deaths, or lending which risks the financial solvency of the lending institution. Some of the regulatory tools available to those regulators can be applied to the use of AI, such as technical standards and independent conformity testing. Justified risk mitigation explanations might be a useful tool for risks which currently arise from human activities in the sector, and where AI will replace that human involvement. For example, we want the AI driving system for an autonomous vehicle to drive as well as, or better than, a human driver, but a driving test for such AIs is hard to devise[649] and might fail to identify

[648] OECD AI Principles 2019, oecd.ai/en/ai-principles.

[649] Thus although the UK Automated Vehicles Bill 2023–24, cl 3(1) empowers the Secretary of State to authorise the use of an automated vehicle on public roads if it satisfies the 'self-driving test', that test is very broadly defined in cl 1(2) as follows:

> A vehicle satisfies the self-driving test if – (a) it is designed or adapted with the intention that a feature of the vehicle will allow it to travel autonomously, and (b) it is capable of doing so, by means of that feature, safely and legally.

Clause 5 permits the Secretary of State to make regulations imposing authorisation requirements, and this would be the route to a generic AI driving test if such a test can eventually be devised.

important defects in the AI's driving. An explanation of the risks which the AI developer foresaw, trained for and tested against could, at least initially, provide a regulator with greater assurance than a driving test.

The second category is where the AI is used to perform a function which was previously undertaken by a human, usually involving a decision, where society expected the human to conform to what I described at the beginning of this chapter as aspirational values. For these values society knows what it wants from the outcome, and can identify whether it was achieved, but cannot define in advance what success or failure might look like. Fairness is a clear example of such an outcome, where the test for whether it has been achieved is that the outcome engenders the emotion of fairness. For these outcomes, as explained in section 8.1, the most appropriate regulatory tool at present is some form of meta-regulation, which I suggest should take the form of justified risk mitigation explanations. Clearly, society will not wish to impose these kinds of regulatory obligations on every AI producer, because that would eventually amount to regulating the whole of human activity. But if the regulatory aim is to reduce the risk that the outcome is not achieved then, as I have already suggested, we can base the imposition of regulation on the seriousness of the consequences if it is not. To revert to a previous example, a recommender system for music listening produces trivial consequences if it is unfair, or inconsistent, or not explainable, and so needs no regulation. By contrast, an inaccurate or inconsistent AI diagnosis for medical treatment has potentially serious consequences, and so regulation is clearly justified.

Over the longer term, our understanding of AI systems which we expect to achieve these aspirational outcomes will improve. As it does so, a wider range of regulatory tools will become available. AI fairness techniques might develop to such a level of sophistication that we can use regulation to mandate particular techniques or combinations of techniques for particular applications – this kind of fairness would not be flexible enough to match the human conception completely, but could be an acceptable compromise to guarantee an overall minimum of fairness in carefully delimited circumstances. It is too early to say where the regulation of AI fairness, and other aspirational values, will end up, but requiring justified risk mitigation explanations where the consequences of failure are potentially serious seems likely to be a useful starting point.

The third and final category is where AI seems to be developing the capacity to undertake some new activity which has no clear parallel or counterpart in current human activity. Here we are unsure whether AI will actually develop this capacity, or if so how it would manifest in outcomes. This means that we can only speculate about the nature of the risks, and thus cannot yet devise effective regulatory safeguards against them. We might describe these as existential or future risks, capturing that uncertainty.

As I write, the UK's November 2023 AI Summit focusing on this category of AI outcomes has concluded, and the US has just published an Executive Order on the

Safe, Secure, and Trustworthy Development and Use of Artificial Intelligence.[650] Policy concerns concentrate on 'frontier AI', described in the UK AI Summit documentation as 'Highly capable general purpose AI models that can perform a wide variety of tasks and match or exceed the capabilities present in today's most advanced models',[651] or on what the US Executive Order terms 'dual-use foundation models'.[652] The outcomes which are of concern to these policy initiatives are those which threaten society as a whole, rather than just individual interests. The UK discussion document for November 2023 AI Summit[653] identifies three broad types of risk:

- societal harms, such as degradation of the information environment, unfairness and bias, disruption of employment;
- misuse risks, such as cybersecurity and disinformation;
- loss of control of AI, through humans handing over control or through AI reducing scope for human control.

The US Executive Order uses the examples of AI which has potential to assist in weapons design, enable cyberattacks or evade human control through deception.

The immediate need is clearly to monitor how these AI systems develop, to identify whether these (or other) risks eventuate and, if possible, how they might be controlled. Both the US and the UK plan to establish AI Safety institutes whose briefs will include to 'carefully examine, evaluate, and test new types of AI so that we understand what each new model is capable of, exploring all the risks, from social harms like bias and misinformation, through to the most extreme risks of all.'[654] These initiatives will bring together experts to monitor and assess AI development and report on risks and potential solutions.

However, experts like these can only be external observers of AI development. It is AI developers themselves who have the first and deepest knowledge about the capabilities of their systems, and who are therefore best placed to identify the risks they create and potential risk mitigations. Section 4.2(a) of the US Executive Order

[650] 30 October 2023, www.whitehouse.gov/briefing-room/presidential-actions/2023/10/30/executive-order-on-the-safe-secure-and-trustworthy-development-and-use-of-artificial-intelligence/.

[651] UK Government, *Safety and Security Risks of Generative Artificial Intelligence to 2025* (October 2023) 2, assets.publishing.service.gov.uk/media/653932db80884d0013f71b15/generative-ai-safety-security-risks-2025-annex-b.pdf).

[652] These are any 'AI model that is trained on broad data; generally uses self-supervision; contains at least tens of billions of parameters; is applicable across a wide range of contexts; and that exhibits, or could be easily modified to exhibit, high levels of performance at tasks that pose a serious risk to security, national economic security, national public health or safety, or any combination of those matters', US Executive Order (n 650), s 3(k).

[653] UK Government Office for Science, *Future Risks of Frontier AI* (October 2023) 4, assets.publishing.service.gov.uk/media/653bc393d10f3500139a6ac5/future-risks-of-frontier-ai-annex-a.pdf); UK Department for Science, Innovation and Technology, *Capabilities and risks from frontier AI* (October 2023), assets.publishing.service.gov.uk/media/65395abae6c968000daa9b25/frontier-ai-capabilities-risks-report.pdf.

[654] UK Prime Minister's speech to the Royal Society (26 October 2023), www.gov.uk/government/speeches/prime-ministers-speech-on-ai-26-october-2023.

recognises this, and places an obligation on the large companies[655] creating dual-use foundation models to report on their training and development processes, including security, who owns and possesses the model and how it is secured, and the results of any internal investigations by the company into flaws and vulnerabilities and their mitigation. This is clearly an obligation to identify risks and explain how they are being addressed, and therefore fits well with the meta-regulatory scheme proposed in chapter eight.

As undesirable outcomes are identified through these processes, and their risks become clearer, regulation will need to require steps to be taken to alleviate those risks. Lawmakers face a choice here. If it is clear what needs to be done, for example, through compliance with cybersecurity standards, then this can simply be mandated. For most of these risks, though, AI producers will have a range of possible mitigation measures they could take, and will thus need to make choices between them.

My argument in respect of fairness was that AI producers are better placed to make good choices than are regulators, provided there is sufficient oversight of their choices. That oversight is provided by requiring them to justify their choices, which allows a regulator to assess the quality of the justification and, if necessary, require stronger risk mitigation. The same argument seems to apply to these broad societal risks – the AI producer is best placed to devise suitable risk mitigation, while a regulator takes on the role of evaluating the producer's plans.

The alternative is to prescribe in regulation how the AI producer must behave. Taking this route assumes that the regulator or lawmaker has sufficient knowledge and expertise to identify the risk and how to manage it. It also assumes that the technology will remain reasonably static, so that the regulatory solution does not need constant updating. Neither of these are true at present, and are unlikely ever to be true. Regulators find it hard to recruit expertise, and AI technology changes far more rapidly than regulation can be amended.

AI's challenges are not merely social – it also challenges our existing models for how we regulate technology. We cannot respond to those challenges in the old ways without risking the very outcomes that we want to avoid.

[655] Companies whose computing power available for such development exceeds the values specified in US Executive Order (n 650), s 4.2(b).

SELECT BIBLIOGRAPHY

Note to readers: Many of the sources in this bibliography are multidisciplinary, or address problems in one discipline from the perspective of a different discipline. I have attempted to list them under a heading which matches what seemed to be the main focus of the piece, but that heading should be treated only as a rough guide.

Social and Psychological Fairness

Alves, Wayne M and Rossi, Peter H, 'Who should get what? Fairness judgments of the distribution of earnings' (1978) 84 *American journal of Sociology* 541

Berger, Joseph, Ridgeway, Cecilia L, Fisek, M Hamit and Norman, Robert Z, 'The legitimation and delegitimation of power and prestige orders' (1998) *American Sociological Review* 379

Greenberg, Jerald, 'A taxonomy of organizational justice theories' (1987) 12 *Academy of Management Review* 9

Ridgeway, Cecilia L, 'Why status matters for inequality' (2014) 79 *American Sociological Review* 1

Thibaut, John W and Walker, Laurens, *Procedural justice: A psychological analysis* (Mahwah NJ, L Erlbaum Associates, 1975)

Tyler, Tom, 'Justice and effective cooperation' (2012) 25 *Social Justice Research* 355

—— 'Procedural Justice', Ch 23 in Austin Sarat (ed), *The Blackwell Companion to Law and Society* (Oxford, Blackwell, 2004) 442–43

Van den Bos, Kees, Maas, Marjolein, Waldring, Ismintha E and Semin, Gün R, 'Toward understanding the psychology of reactions to perceived fairness: The role of affect intensity' (2003) 16 *Social Justice Research* 151

Wilson, Bart J, 'Contra private fairness' (2012) 71 *American Journal of Economics and Sociology* 407

Philosophical and Economic Fairness

Arneson, Richard, 'Equality and Equality of Opportunity for Welfare' (1989) 55 *Philosophical Studies* 79

—— 'Four conceptions of equal opportunity' (2018) 128 *The Economic Journal* F152

Cohen, GA, 'On the Currency of Egalitarian Justice' (1898) 99 *Ethics* 906

Dworkin, Ronald, 'What is Equality? Part 1: Equality of Welfare' (1981) 10 *Philosophy & Public Affairs* 185

—— 'What is Equality? Part 2: Equality of Resources' (1981) 10 *Philosophy & Public Affairs* 283

Fleurbaey, Marc, 'Equal opportunity or equal social outcome?' (1995) 11 *Economics & Philosophy* 25

Floridi, Luciano, Cowls, Josh, King, Thomas C and Taddeo, Mariarosaria, 'How to Design AI for Social Good: Seven Essential Factors' (2020) 26 *Science and Engineering Ethics* 1771

Gintis, Herbert, 'Behavioral Ethics Meets Natural Justice' (2006) 5 *Politics, Philosophy and Economics* 5

Heidari, Hoda, Ferrari, Claudio, Gummadi, Krishna and Krause, Andreas, 'Fairness behind a veil of ignorance: A welfare analysis for automated decision making' (2018) 31 *Advances in Neural Information Processing Systems*

Heidari, Hoda, Loi, Michele, Gummadi, Krishna P and Krause, Andreas, 'A moral framework for understanding fair ML through economic models of equality of opportunity' (2019) *Proceedings of the Conference on Fairness, Accountability, and Transparency* 181

High Level Expert Group on Artificial Intelligence, *Ethics Guidelines for Trustworthy AI* (European Commission, 2019), digital-strategy.ec.europa.eu/en/library/ethics-guidelines-trustworthy-ai

John-Mathews, Jean-Marie, Cardon, Dominique and Balagué, Christine, 'From reality to world. A critical perspective on AI fairness' (2022) 178 *Journal of Business Ethics* 945

Lee, Michelle Seng Ah, Floridi, Luciano and Singh, Jatinder, 'Formalising trade-offs beyond algorithmic fairness: lessons from ethical philosophy and welfare economics' (2021) 1 *AI and Ethics* 529

Leslie, David, *Understanding artificial intelligence ethics and safety: A guide for the responsible design and implementation of AI systems in the public sector* (Alan Turing Institute 2019), www.turing.ac.uk/sites/default/files/2019-06/understandingartificialintelligenceethicsandsafety.pdf

Mittelstadt, Brent, 'Principles alone cannot guarantee ethical AI' (2019) 1 *Nature Machine Intelligence* 501

Mullainathan, Sendhil, 'Algorithmic Fairness and the Social Welfare Function' (2018) *ACM EC'18: Proceedings of the 2018 ACM Conference on Economics and Computation*

Rabin, Matthew, 'Incorporating fairness into game theory and economics' (1993) 83 *The American Economic Review* 1281

Rambachan, Ashesh, Kleinberg, Jon, Ludwig, Jens and Mullainathan, Sendhil, 'An economic perspective on algorithmic fairness' (2020) 110 *AEA Papers and Proceedings* 91

Rawls, John, *A Theory of Justice* (Cambridge MA, Harvard University Press, 1971)

Roemer, John, 'Equality of opportunity: A progress report' (2002) 19 *Social Choice and Welfare* 455

Roemer, John E and Trannoy, Alain, 'Equality of opportunity: Theory and measurement' (2016) 54 *Journal of Economic Literature* 1288

Sobel, Joel, 'Interdependent preferences and reciprocity' (2005) 42 *Journal of Economic Literature* 392

Whittlestone, Jess, Nyrup, Rune, Alexandrova, Anna, Dihal, Kanta and Cave, Stephen, *Ethical and societal implications of algorithms, data, and artificial intelligence: a roadmap for research* (Nuffield Foundation and Leverhulme Centre for the Future of Intelligence, University of Cambridge, 2019), www.nuffieldfoundation.org/wp-content/uploads/2019/12/Ethical-and-Societal-Implications-of-Data-and-AI-report-Nuffield-Foundat.pdf

Winter, Fabian, 'Fairness norms can explain the emergence of specific cooperation norms in the battle of the Prisoner's Dilemma' (2014) 38 *Journal of Mathematical Sociology* 302

Technological Fairness

Binns, Reuben, 'On the apparent conflict between individual and group fairness' (2020) *Proceedings of the 2020 Conference on Fairness, Accountability, and Transparency* 514

Carey, Alycia N and Wu, Xintao, 'The Fairness Field Guide: Perspectives From Social And Formal Sciences' (2022) *AI and Ethics* 1

Chen, Jiahao, Kallus, Nathan, Mao, Xiaojie, Svacha, Geoffry and Udell, Madeleine, 'Fairness under unawareness: Assessing disparity when protected class is unobserved' (2019) *Proceedings of the Conference on Fairness, Accountability, and Transparency* 339

Chockler, Hana and Halpern, Joseph Y, 'On testing for discrimination using causal models' (2022) 36 *Proceedings of the AAAI Conference on Artificial Intelligence* 5548

Corbett-Davies, Sam, Gaebler, Johann, Nilforoshan, Hamed, Shroff, Ravi and Goel, Sharad, 'The measure and mismeasure of fairness' (2023) *arXiv preprint arXiv:1808.00023*

Dwork, Cynthia, Hardt, Moritz, Pitassi, Toniann, Reingold, Omer and Zemel, Richard, 'Fairness through awareness' (2012) *Proceedings of the 3rd innovations in theoretical computer science conference, ACM* 214

Friedler, Sorelle A, Scheidegger, Carlos and Venkatasubramanian, Suresh, 'The (im)possibility of fairness: Different value systems require different mechanisms for fair decision making' (2021) 64 *Communications of the ACM* 136

Jacobs, Abigail Z and Wallach, Hanna, 'Measurement and fairness' (2021) *Proceedings of the 2021 ACM Conference on Fairness, Accountability, And Transparency* 375

John-Mathews, Jean-Marie, Cardon, Dominique and Balagué, Christine, 'From reality to world. A critical perspective on AI fairness' (2022) 178 *Journal of Business Ethics* 945

Kasy, Maximilian and Abebe, Rediet, 'Fairness, equality, and power in algorithmic decision-making' (2021) *Proceedings of the 2021 ACM Conference on Fairness, Accountability, and Transparency. FAccT '21*, 576

Lewicki, Kornel, Seng Ah Lee, Michelle, Cobbe, Jennifer and Singh, Jatinder, 'Out of Context: Investigating the Bias and Fairness Concerns of "Artificial Intelligence as a Service"' (2023) *CHI '23: ACM Human Factors in Computing*

Mitchell, Shira, Potash, Eric, Barocas, Solon, D'Amour, Alexander and Lum, Kristian, 'Algorithmic fairness: Choices, assumptions, and definitions' (2021) 8 *Annual Review of Statistics and Its Application* 141

Selbst, Andrew D, Boyd, Danah, Friedler, Sorelle A, Venkatasubramanian, Suresh and Vertesi, Janet, 'Fairness and abstraction in sociotechnical systems' (2019) *Proceedings of the Conference on Fairness, Accountability, and Transparency* 59

Veale, Michael, Van Kleek, Max and Binns, Reuben, 'Fairness and accountability design needs for algorithmic support in high-stakes public sector decision-making' (2018) *Proceedings of the 2018 CHI Conference On Human Factors in Computing Systems* 1

Verma, Sahil and Rubin, Julia, 'Fairness Definitions Explained', in *Fair-Ware'18*: IEEE/ACM International Workshop on Software Fairness, Gothenburg (29 May 2018)

Legal Fairness

Barocas, Solon and Selbst, Andrew D, 'Big data's disparate impact' (2016) 104 *California Law Review* 671

Biega, Asia and Finck, Michèle, 'Reviving purpose limitation and data minimisation in data-driven systems' (2021) *Technology and Regulation* 44

Clifford, Damian and Ausloos, Jef, 'Data protection and the role of fairness' (2018) 37 *Yearbook of European Law* 130

de Vries, Katja, 'Transparent Dreams (Are Made of This): Counterfactuals as Transparency Tools in ADM' (2021) 8 *Critical Analysis of Law* 121

European Data Protection Supervisor, 'Opinion on Coherent Enforcement of Fundamental Rights in the Age of Big Data' (EDPS 2016) Opinion 8/201

Georgosouli, Andromachi, 'Metarules, Judgment and the Algorithmic Future of Financial Regulation in the UK' (2023) 43 *Oxford Journal of Legal Studies* 777

—— 'The FSA's "Treating Customers Fairly" (TCF) initiative: What is so good about it and why it may not work' (2011) 38 *Journal of Law and Society* 405

Hacker, Philipp, 'Teaching fairness to artificial intelligence: existing and novel strategies against algorithmic discrimination under EU law' (2018) 55 *Common Market Law Review* 1143

ICO Guidance, 'What do we need to do to ensure lawfulness, fairness, and transparency in AI systems?', ico.org.uk/for-organisations/guide-to-data-protection/key-dp-themes/guidance-on-ai-and-data-protection/how-do-we-ensure-fairness-in-ai

ICO, 'Human bias and discrimination in AI systems', ico.org.uk/about-the-ico/media-centre/ai-blog-human-bias-and-discrimination-in-ai-systems

Malgieri, Gianclaudio, 'The concept of fairness in the GDPR: a linguistic and contextual interpretation' (2020) *Proceedings of the 2020 Conference on fairness, accountability, and transparency* 154

Oswald, Marion, 'Algorithm-assisted decision-making in the public sector: framing the issues using administrative law rules governing discretionary power' (2018) 376 *Philosophical Transactions of the Royal Society A* 20170359

Singh, Jatinder, Walden, Ian, Crowcroft, Jon and Bacon, Jean, 'Responsibility & Machine Learning: Part of a Process' (2016), papers.ssrn.com/sol3/papers.cfm?abstractid=2860048

Vale, Daniel, El-Sharif, Ali and Ali, Muhammed, 'Explainable artificial intelligence (XAI) post-hoc explainability methods: Risks and limitations in non-discrimination law' (2022) 2 *AI and Ethics* 815

Wachter, Sandra, Mittelstadt, Brent and Russell, Chris, 'Counterfactual Explanations Without Opening the Black Box: Automated Decisions and the GDPR' (2018) 31 *Harvard Journal of Law & Technology* 841

—— 'Why fairness cannot be automated: Bridging the gap between EU non-discrimination law and AI' (2021) 41 *Computer Law & Security Review* 105567

AI Regulation

Black, Julia and Murray, Andrew, 'Regulating AI and machine learning: setting the regulatory agenda' (29) *European Journal of Law and Technology*, 11

Braithwaite, John, 'Enforced self-regulation: a new strategy for corporate crime control' (1982) 80 *Michigan Law Review* 1466

Burr, Christopher and Leslie, David, 'Ethical assurance: a practical approach to the responsible design, development, and deployment of data-driven technologies' (2023) 3 *AI and Ethics* 73

Cobbe, Jennifer, Veale, Michael and Singh, Jatinder, 'Understanding accountability in algorithmic supply chains' (2023) *Proceedings of the 2023 ACM Conference on Fairness, Accountability, and Transparency* 1186

Edwards, Lilian and Veale, Michael, 'Enslaving the algorithm: From a 'right to an explanation to a 'right to better decisions?' (2018) 16 *IEEE Security & Privacy* 46

Gervais, Daniel J, 'Towards an effective transnational regulation of AI' (2021) *AI & Society* 1

Grabowsky, Peter, 'Meta-regulation' in Peter Drahos (ed), *Regulatory theory: foundations and applications* (Canberra, ANU Press, 2017) ch 9

Petit, Nicolas and De Cooman, Jerome, 'Models of law and regulation for AI' in Anthony Elliott (ed), *The Routledge Social Science Handbook of AI* (London, Routledge, 2021) ch 13

Reed, Chris, 'How should we regulate artificial intelligence?' (2018) *Philosophical Transactions of the Royal Society A* 376

Reed, Chris, Grieman, Keri and Early, Joseph, 'Non-Asimov explanations – regulating AI through transparency' in *Law in the Era of Artificial Intelligence: Nordic Yearbook of Law & Informatics 2020–2021*, ch 14, irilaw.org/2022/02/16/new-publication-nordic-yearbook-of-law-and-informatics-2020-2021.

Scherer, Matthew U, 'Regulating Artificial Intelligence Systems: Risks, Challenges, Competencies, and Strategies' (2016) 29 *Harvard Journal of Law & Technology* 353

Scott, Colin, 'Regulating everything: From mega-to meta-regulation' (2012) 60 *Administration* 57

Singapore Infocomm Media Development Authority (IMDA), *Fact Sheet – Open-Sourcing of AI Verify and Set Up of AI Verify Foundation* (2023), www.imda.gov.sg/-/media/imda/files/news-and-events/media-room/media-releases/2023/06/7-jun---ai-annoucements---annex-a.pdf

Singapore PDPC, *Model Artificial Intelligence Governance Framework*, 2nd edn (2020), www.pdpc.gov.sg/-/media/files/pdpc/pdf-files/resource-for-organisation/ai/sgmodelaigovframework2.pdf

UK Government Office for Science, *Future Risks of Frontier AI* (October 2023), assets.publishing.service.gov.uk/media/653bc393d10f3500139a6ac5/future-risks-of-frontier-ai-annex-a.pdf

UK Department for Science, Innovation and Technology, *Capabilities and risks from frontier AI* (October 2023), assets.publishing.service.gov.uk/media/65395abae6c968000daa9b25/frontier-ai-capabilities-risks-report.pdf

UK House of Lords Artificial Intelligence Committee, *AI in the UK: ready, willing and able?* (2018) HL Paper 100

UK White Paper, *A pro-innovation approach to AI regulation* (2023), www.gov.uk/government/publications/ai-regulation-a-pro-innovation-approach/white-paper

US White Paper, Executive Office of the President, *Artificial intelligence, automation and the economy* (2016)

Viljanen, Mika and Parviainen, Henni, 'AI applications and regulation: Mapping the regulatory strata' (2022) 3 *Frontiers in Computer Science*, article 779957

AI Trustworthiness, Consistency and Explainability

Burkart, Nadia and Huber, Marco F, 'A survey on the explainability of supervised machine learning' (2021) 70 *Journal of Artificial Intelligence Research* 245

Doran, Derek, Schulz, Sarah and Besold, Tarek R, 'What does explainable AI really mean? A new conceptualization of perspectives' (2018) *CEUR Workshop Proceedings*, vol 2071

Fabris, Adriano, 'Can we trust machines? The role of trust in technological environments' in Adriano Fabris (ed), *Trust: A Philosophical Approach* (Cham, Springer, 2020) 123

Gundersen, Odd Erik and Kjensmo, Sigbjørn 'State of the art: Reproducibility in artificial intelligence' (2018) 32 *Proceedings of the AAAI Conference on Artificial Intelligence* 1644

Haibe-Kains, Benjamin et al, 'Transparency and reproducibility in artificial intelligence' (2020) 586 *Nature* E14

Li, Bo, Qi, Peng, Liu, Bo, Di, Shuai, Liu, Jingen, Pei, Jiquan, Yi, Jinfeng and Zhou, Bowen, 'Trustworthy AI: From principles to practices' (2023) 55 *ACM Computing Surveys* 1

Pineau, Joelle, Vincent-Lamarre, Philippe, Sinha, Koustuv, Larivière, Vincent, Beygelzimer, Alina, d'Alché-Buc, Florence, Fox, Emily and Larochelle, Hugo, 'Improving reproducibility in machine learning research (a report from the NeurIPS 2019 reproducibility program)' (2021) 22 *Journal of Machine Learning Research* 7459

Rutter, Jason, 'From the sociology of trust towards a sociology of "e-trust"' (2001) 2 *International Journal of New Product Development & Innovation Management* 371

Shumailov, Ilia, Shumaylov, Zakhar, Zhao, Yiren, Gal, Yarin, Papernot, Nicolas and Anderson, Ross, 'The Curse of Recursion: Training on Generated Data Makes Models Forget' (2023) *arXiv preprint arXiv:2305.17493*

Generative AI and Copyright

Craig, Carys J, 'The AI-copyright challenge: Tech-neutrality, authorship, and the public interest' in Ryan Abbott (ed), *Research Handbook on Intellectual Property and Artificial Intelligence* (Cheltenham, Edward Elgar, 2022) ch 7

Franceschelli, Giorgio and Musolesi, Mirco, 'Copyright in generative deep learning' (2022) 4 *Data & Policy* e17

Gillotte, Jessica L, 'Copyright Infringement in AI-Generated Artworks' (2020) 53 *UC Davis Law Review* 2655

Guadamuz, Andrés, 'A Scanner Darkly: Copyright Infringement in Artificial Intelligence Inputs and Outputs' (2023) papers.ssrn.com/sol3/papers.cfm?abstractid=437120

Samuelson, Pamela, 'Legal Challenges to Generative AI, Part I' (2023) 66 *Communications of the ACM* 20

Sunray, Eric, 'Sounds of Science: Copyright Infringement in AI Music Generator Outputs' (2021) 29 *Catholic University Journal of Law & Technology* 185

INDEX

www.ingramcontent.com/pod-product-compliance
Ingram Content Group UK Ltd.
Pitfield, Milton Keynes, MK11 3LW, UK
UKHW020808290926
15470UKWH00003B/1026

9 781509 976805